A guide to manual materials handling

Dedication

To our families and colleagues.

A guide to manual materials handling

A. Mital

University of Cincinnati, USA

A. S. Nicholson

Hu-Tech Associates, UK

M. M. Ayoub

Texas Tech University, USA

Taylor & Francis
London • Washington, DC
1997

UK Taylor & Francis Ltd, 4 John St., London WC1N 2ET
USA Taylor & Francis Inc., 1900 Frost Road, Suite 101, Bristol, PA 19007

British Library Cataloguing in Publication Data
A catalogue record for this book is available from the British Library.

ISBN 0–7484–0728–6

Library of Congress Cataloging-in-Publication Data are available

Cover design by Amanda Barragry.

Phototypesetting by Solidus Graphic Communication, Bristol.

Printed in Great Britain by Athenæum Press Ltd., Gateshead.

Contents

Dedication		ii
Preface to the Second Edition		ix
Preface to the First Edition		xi
PART 1		
Chapter 1.	**The world-wide scope of manual materials handling**	3
	Introduction	3
	Scope of this guide	3
	Available statistics of various countries	3
	Legislation of various countries addressing the problem of MMH	5
	Conclusions	13
	References	13
Chapter 2.	**Factors to be considered in designing manual materials handling tasks**	14
	Introduction	14
	Physique/anthropometry/strength	14
	Physical fitness/spinal mobility	15
	Age/gender differences	15
	Psychophysical factors/motivation	17
	Training and selection	17
	Effects of static work	19
	Posture/handling techniques	19
	Load characteristics	20
	Handles/coupling	21
	Repetitive handling	22
	Asymmetrical lifting/load asymmetry	22
	Confined environments/spatial restraints	23
	Safety aspects	23
	Protective equipment	24
	Handling in a hot environment	24
	Task duration	24
	Work organization	25
	References	25
Chapter 3.	**Design approaches to solving manual materials handling problems**	28
	Introduction	28
	The epidemiological approach	28
	The design criteria	33
	The biomechanical approach	34
	The design criteria	35
	The physiological approach	38
	The design criterion	39
	The psychophysical approach	41
	The design criteria	42
	Comparison of design approaches	42
	Margin of safety for the human back	48
	Epidemiological approach	48

Biomechanical approach 49
Psychophysical approach 53
Physiological approach 54
Conclusion 54
References 55

PART II
Chapter 4. **Lifting** 61
Two-handed lifting 61
The design criteria 61
Design database 62
Working duration multiplier 62
Limited headroom multiplier 67
Asymmetrical lifting multiplier 67
Load asymmetry multiplier 67
Couplings multiplier 68
Load placement clearance multiplier 68
Heat stress multiplier 69
Use of tables 69
One-handed lifting 69
Two-person lifting 71

Chapter 5. **Pushing** 72
Two-handed pushing 74
One-handed pushing 74

Chapter 6. **Pulling** 78
Two-handed pulling 82
One-handed pulling 83

Chapter 7. **Carrying** 84
Two-handed carrying 86
One-handed carrying 86

Chapter 8. **Holding** 87
Reference 90

Chapter 9. **Materials handling in unusual postures** 91
Lifting 91
Pushing 102
Pulling 106
References 106

Chapter 10. **Designing and evaluating multiple-task manual materials handling jobs: using this guide** 107
Introduction 107
Design/evaluation procedure 107
Case study 1 108
Case study 2 111
Case study 3 113
Summary 115
References 115

Chapter 11. **High and very high frequency manual lifting/lowering and carrying (load transfer)** 116
Introduction 116
Weight limits for high and very high frequency manual handling tasks 116
Lifting and lowering 116
Carrying and turning (load transfer) 119
Endurance time at high lifting frequencies 121
References 121

Chapter 12. **Determination of rest allowances** 122
Introduction 122
Types of allowances 122
Rest allowance determination procedure 122
Examples 126
References 127
Rest allowance determination program source listing 128

Chapter 13. **Mechanical aids** 132

Index 138

Preface to the Second Edition

The global popularity of the first edition, the excellent reviews it received in professional journals and magazines, comments from the participants of workshops we have conducted in different parts of the world, and complimentary remarks from our colleagues in many countries motivated us to undertake the preparation of the second edition of *A Guide to Manual Materials Handling*. Strong motivation also came from the fact that back injuries remain a cause for serious concern, at least in the United States, in terms of number, severity, and cost.

Recent data from several public agencies (e.g. Bureau of Labor Statistics and the National Safety Council) indicate that most injuries occur in manufacturing settings, having overtaken or equalled injuries in the construction industry. In 1993, for the first time, the incidence rate for nonfatal cases without lost workdays in the manufacturing industry exceeded the corresponding incidence rate in the construction industry. Also, for the first time, in 1994, the incidence rate for nonfatal lost workday cases in the manufacturing industry equalled with the corresponding incidence rate in the construction industry. Since the back is involved more often than any other part of the body (see Chapter 1 for data), one can logically conclude that back injuries are most prevalent in the manufacturing industry.

Despite information on injuries, the exact magnitude of the cost of back injuries still remains a matter of speculation, at least in the United States. A 1996 report by the National Institute for Occupational Safety and Health estimates the annual cost attributable to back injuries to be between 50 and 100 billion dollars – an underestimation or overestimation ranging from 50% to 100%. According to the National Safety Council, between 1972 and 1994 the direct annual cost of all injuries in the United States quadrupled to $120.4 billion. Taking a very conservative estimate that back injuries are no more than 25% of all injuries, and applying the health care industry indirect cost estimate (indirect costs are 4.5 times the direct costs), the back injury problem turns out to cost over $150 billion per year.

This second edition of the *Guide* includes three new chapters and Chapter 1 has been rewritten. Chapter 10 describes a procedure for designing and evaluating multiple-task materials handling jobs. It shows, with the help of three case studies, how this guide may be used. In Chapter 11, we present weight and endurance time limits for lifting, lowering, and carrying tasks performed at rates of 16 times per minute or higher. Chapter 12 provides a procedure for determining rest allowances associated with performing manual materials handling tasks. This chapter also includes software, written in Basic, to calculate rest allowances. (Software packages to analyse multiple-task manual materials handling jobs, and to estimate spinal compressive/shear forces and associated margins of safety are available from the principal author at additional cost.)

Chapters 2–9, and the old Chapter 10 (Chapter 13 in the second edition) remain, more or less, untouched; some minor corrections, additions, and deletions have been made.

It should be noted that the design data provided in Chapters 4–9 and Chapter 11 have come from a variety of sources, including the University of Cincinnati, Liberty Mutual Insurance Company, the University of Michigan, the University of Surrey, and Texas Tech University. Since the sources of data are numerous, and in most cases the data are not reproduced in the original form, it is not possible to mention them all by name.

Finally, we hope that the second edition of this guide will be as useful and valuable as the first edition, and will play a significant role in containing the number, severity, and cost of back injuries.

Anil Mital
Andrew Nicholson
Moh Ayoub

May 1997

Preface to the First Edition

Almost every occupational setting requires some form of material handling. Space limitations, varied nature of the activity, and the reluctance to make substantial investment in automated equipment are some of the reasons favouring manual handling of materials. Invariably, the abilities of individuals to perform these activities, either frequently or occasionally, are exceeded, resulting in severe chronic or acute injuries.

The direct cost of manual handling injuries has now risen to approximately 15 billion dollars annually in the United States. The indirect costs may be as much as four times this amount. Many other countries, for instance the United Kingdom and Australia, also find the severity and cost of manual materials handling related injuries unacceptable and a cause for serious concern.

The need to control the cost and severity of injuries caused by handling different kinds of materials has led to several concerted efforts to develop guidelines or manuals for designing these jobs. The most notable of these guidelines are the 1981 Work Practices Guide for Manual Lifting developed by the National Institute for Occupational Safety and Health (NIOSH), the 1980 Force Limits in Manual Work developed by the Robens Institute of the University of Surrey, the 1991 revision of NIOSH's 1981 Work Practices Guide to Manual Lifting, and the most recent draft on manual lifting from the Health and Safety Executive, United Kingdom. These guidelines are, however, mostly limited to manual lifting activities. Many manual materials handling activities involve pushing, pulling, carrying or holding, or are a combination of several of these. The complicated and diverse nature of manual materials handling activities almost mandates us to consider all possible forms of manual materials handling activities in a guide. This guide is our response to fill the existing gap.

Our goal has not only been to put forth a comprehensive guide to designing manual materials handling activities, but also to reconcile the differences between the various design approaches. This guide, therefore, considers the epidemiological approach, the biomechanical approach, the physiological approach, and the psychophysical approach to manual materials handling job design. Furthermore, instead of trying to provide a simple formula that may be severely limited, or totally inadequate, in solving a complex occupational problem, as some guides have done, we have chosen to provide detailed design data tables that express the population's abilities to perform manual materials handling activities, safely. The choice of the final form in which these data are used (expert systems, percentile equations, etc.) has been left to the user.

This guide, written in simple language, has been organized in two main parts. Part I reviews legislation and limits that different countries have imposed on manual materials handling, factors that affect populations' abilities to perform manual materials handling activities, and the various design approaches and design criteria used in the development of this guide. Major references used in the development of the guide are given at the end of each chapter.

The second part of this guide provides a manual materials handling design database in six separate chapters. In addition to basic materials handling activities, such as lifting, pushing, pulling, and carrying, holding and materials handling in unusual postures are also considered. The last chapter shows the numerous devices that are available to aid manual materials handling.

It is our hope that this guide will be helpful to practitioners in the field to solve manual materials handling problems and, in general, to control the rate and severity of injury for these activities.

Anil Mital
Andrew Nicholson
Moh Ayoub

September 1992

Part I

Chapter 1

The world-wide scope of manual materials handling

Introduction

Manual materials handling (MMH) creates special problems for many different workers world-wide. Labourers engaged in jobs which require lifting/lowering, carrying, and pushing/pulling of heavy materials have increased rates of musculoskeletal injuries, especially to the back. The world community is beginning to recognize the problems MMH presents for workers. It is necessary to look at the scope of MMH problems to get some idea of the true size of the overall problem. In the following sections, after a review of some of the statistics available, the legislative actions of various countries with respect to MMH are addressed.

Scope of this guide

This guide is intended to include all activities involved in MMH (lifting, pushing, pulling, carrying, and holding) as opposed to other existing guides that focus only on manual lifting. The recommendations are provided in the form of design data that can be used to design different MMH activities.

The guide is divided into two parts. Part I outlines the scope of the problem, discusses the factors that influence a person's capacity to perform MMH activities and/or that should be modified to reduce the risk of injuries, and reviews the various design approaches to solve the MMH problem. Part II provides specific design data in seven distinct chapters (Chapters 4–9, and 11) for six different MMH areas. Chapter 10 describes a generalized procedure for designing and evaluating multiple-activity MMH jobs. In Chapter 12, a procedure for determining rest allowances is described. The last chapter of the guide describes various mechanical devices that are available to aid MMH activities.

Available statistics of various countries

While most countries acknowledge the problem MMH creates for workers, accurate statistics of injuries attributable to MMH are still hard to find. According to Sweden's National Board of Occupational Safety and Health (1983), the most common types of injury to workers in that country are strains and sprains due to overexertion of body parts. These injuries due to overexertion account for almost 18% of injuries, while another 12% of personal injuries to workers are skeletal. When injuries are classified by occupation, the category 'materials, goods, and packaging' accounts for more than 23% of injuries, by far the largest proportion of any one segment of the labour force (National Board of Occupational Safety and Health, 1983).

Other countries report similar findings. In 1982, the Health and Safety Commission reported over 70000 injuries due to MMH in the UK. During the financial year 1986–87 there were at least 110000 manual handling injuries, half of which resulted in more than 3 days' absence from work (Health and Safety Commission, 1988). In 1988–89, 27.5% of reported accidents in the UK involved manual handling. The cost of these accidents, including lost output, medical treatment, and individual suffering, exceeded £90 million. The total annual cost of musculoskeletal disease exceeds £25 billion (Health and Safety Commission, 1991). Rawling and O'Halloran (1988) report that Australia's State Electricity Commission of Victoria (SECV) estimates that 34% of compensation costs are paid to workers suffering injuries due to MMH. In Luxembourg, data from the Association d'Assurance contre les Accidents showed that acute back disorders accounted for 2% of occupational accidents reported (or 286 out of 15559 reported). Of these, 181 cases (or 78%) of back disorders involved lifting/lowering and carrying and another 26 cases (another 11%) resulted from pushing/pulling loads (Metzler, 1985). Figures 1.1 and 1.2 show the relationship between the number of injuries and their cost from 1972 to 1994 and the magnitude of back injuries in relation to injuries in other parts of the body from 1987 to 1994, respectively, in the USA. It is clear from these figures that not only

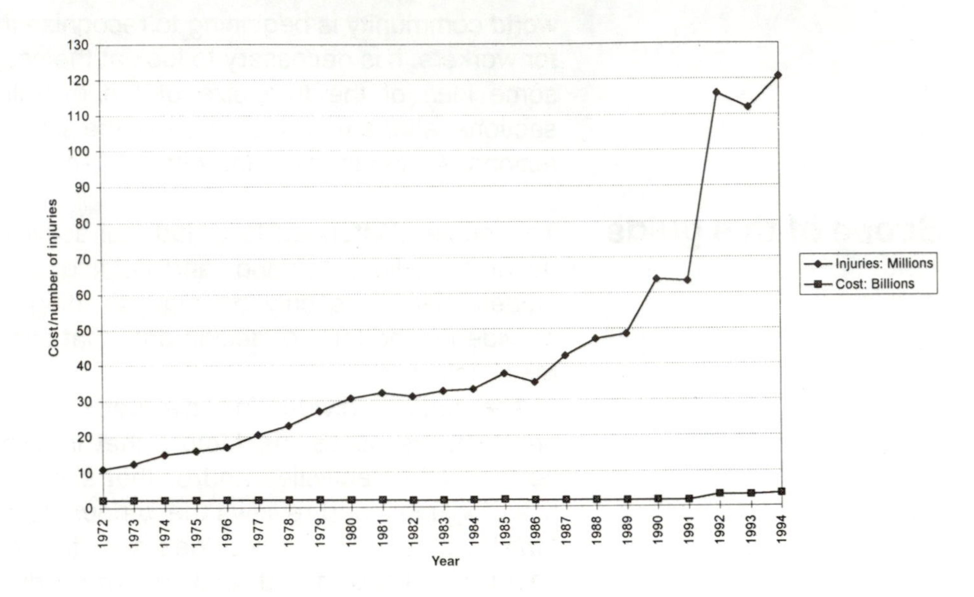

Figure 1.1 Number and cost of work injuries in the USA. Compiled from data published by the National Safety Council (1972–1994).

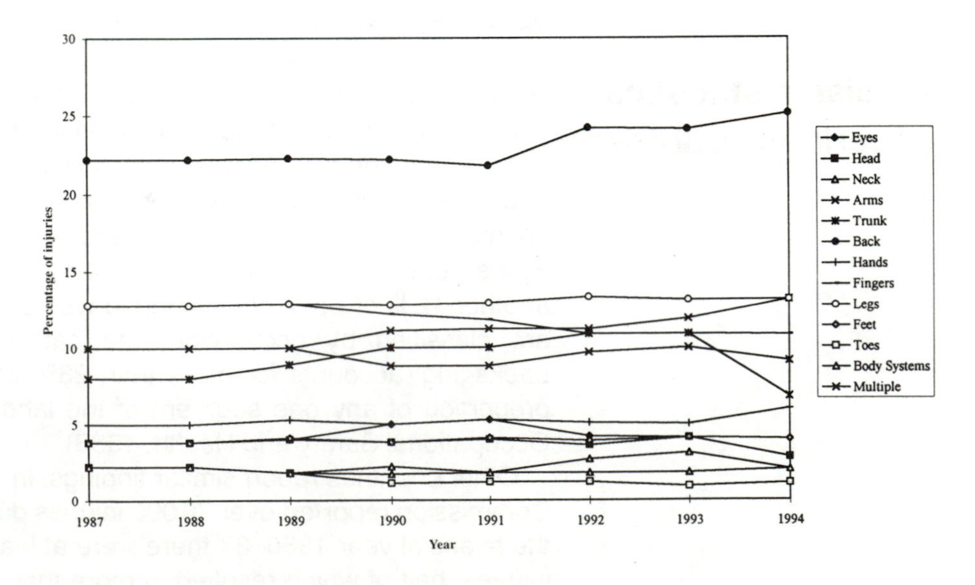

Figure 1.2 Proportion of injuries by part of body injured. Compiled by the National Safety Council from data from various state Labor Departments, 1987–1994.

are injuries to the back more frequent than to any other body part, but the rate of rise in cost is also very high, especially in the last few years. Manual handling has been known to be one of the prime causes of back injury. It is reported in the USA that 60% of people suffering from lower-back injuries claim overexertion caused the injury. Furthermore, back injuries result in much more lost job time (Chaffin, 1987). It has been reported by the National Institute of Occupational Safety and Health in a recent report on the National Occupational Research Agenda (NIOSH, 1996) that the total cost of back injuries in the USA in 1991 was between 50 and 100 billion dollars.

Many more countries lack data on the magnitude of occupational injuries caused by MMH. For practical purposes, there are no regular statistics on MMH injury in the Netherlands (Zuidema, 1985), Germany (Hettinger, 1985), or Denmark (Biering-Sorensen, 1985).

Legislation of various countries addressing the problems of MMH

The seemingly poor state of gathering statistics and reporting information on MMH injuries, however, has not led to the denial of the problem nor inhibited legislative action on the part of several governments. Table 1.1 shows countries that are currently known to have some type of government-established guidelines or laws dealing with MMH. Table 1.2 shows additional countries with current guidelines, advisory standards or legislation addressing MMH. Table 1.3 lists the limits on weight to be lifted and carried in different countries.

NIOSH has addressed the concern for industrial lower-back injuries by developing a set of guidelines followed in the USA. The NIOSH guidelines are advisory in nature but they are generally adhered to by US industries. The NIOSH guidelines do not provide a rule for lifting capacity but they do indicate safe limits. The so-called safe limits tend to be slightly lower than capacity, but the level of protection they provide is unknown.

The NIOSH *Work Practices Guide to Manual Lifting* equation (1981) was revised in 1991 (Waters *et al.*, 1993). The new revised equation utilizes Recommended Weight Limit (RWL) instead of the Action Limit (AL) and the Maximum Permissible Limit (MPL) of the 1981 guide. The RWL is based on three criteria: the biomechanical criterion, the physiological criterion, and the psychophysical criterion.

The Recommended Weight Limit is defined as:

$$\mathrm{RWL} = 23\,(25/H)(1-(0.003\,|\,V-75\,|))(0.82+4.5/D)(1-0.0032A)(FM)(CM)\ \mathrm{kg}$$

where H = horizontal location (cm) forward of midpoint between ankles at origin or destination of lift
V = vertical location (cm) at origin of lift
D = vertical travel distance (cm) between origin and destination of lift
A = angle of twist in degrees
FM = frequency multiplier (see Table 1.4)
CM = coupling multiplier (see Table 1.5)

Australia has several advisory standards in reference to MMH. The Australian Council of Trade Unions (ACTU) has a standard which provides guidelines for manual handling (ACTU, 1983). This standard suggests that (1) no worker be required to handle a load exceeding 16 kg unaided, and (2) no worker be required to handle packages, bags or loads exceeding 240 kg over a 15 min period. It is interesting to note

Table 1.1 General Provisions

Austria	'Workers may be allocated to lifting, carrying and transport of loads only in accordance with their constitution and physical strength' (1983)
Canada	Material handling regulations under the Labour Code state: 'Where, because of the weight, size, shape, toxicity or other characteristics of a material or object, the manual handling of that material or object may endanger the safety or health of an employee, the employer of any such employee shall ensure to the extent that is reasonably practicable, that the material or object is not handled manually'
	In Quebec, Canada, at the level of the provincial jurisdiction, the Safety Code for the Construction Industry states: 'Mechanical apparatus shall be provided and used for carrying material when the safety of the worker is jeopardised'
Guatemala	General regulations on Occupational Hygiene and Safety provide that loads transported by workers shall be suited to their physical powers, taking into account the character, weight and volume of the load and the distance and road to be travelled
Ireland	The Factory Act provides that no person shall be required to lift, carry or move a load so heavy that it may cause him injury
New Zealand	The Factories and Commercial Premises Act, 1981, section 26 on carrying of heavy loads, states: 'No worker in an undertaking shall be required to lift or carry a load which is so heavy that it could injure him (her)'
Pakistan	According to the Factories Act: 'No person shall be employed in any factory to lift, carry or move any load so heavy as to be likely to cause him injury'
Sri Lanka	The Factories Ordinance prescribes that no person shall be employed to lift, carry or move any load as heavy as to be likely to cause injury to him
Egypt, Malaysia, Netherlands, Austria, Luxembourg, Hungary, and Cameroon have general provisions which apply to women, children, and/or younger workers	

(Source: ILO, 1988)

that the ACTU guide also includes NIOSH manual lifting guidelines in an appendix as a reference.

In the state of Victoria, Australia, an Occupational Health and Safety Act set forth regulations (Department of Labour, 1988) with the objective of reducing MMH injuries and their severity. The act requires employers to assess and control risks associated with MMH activities on the job. The regulation provides a number of factors for employers to assess risk to employees. If a risk has been assessed, the employer must redesign the manual handling task to remove or reduce the risk, even to the point of including the introduction of mechanical aids and training.

The UK has many acts addressing MMH in a more or less direct manner (David, 1985). Among them are the following: the Mines and Quarries Act 1954, Section 93; Agriculture (Safety, Health and Welfare Provisions) Act 1956, Section 2; Factories Act 1961; and the Offices, Shops and Railway Premises Act 1963. None of these acts, however, sets any kind of limits or presents guidelines for MMH. In 1991, the Health and Safety Commission drafted proposals for regulations and guidance for handling loads at work. This consultative document sets out the main principles to follow in designing work to minimize handling

Table 1.2 Additional countries with current guidelines, advisory standards or legislation addressing manual materials handling

The USA has guidelines for the maximum acceptable weight of compression on the spine, set forth in the *Revised NIOSH Lifting Equation* developed by the National Institute for Occupational Safety and Health (NIOSH; Waters *et al.*, 1993). The guidelines are advisory in nature

The UK has several legislative acts which require the reporting of manual handling accidents and lumbar injuries to the Health and Safety Executive

In Japan, the maximum weight of lift is set forth by law

The Netherlands has two legislative acts which ensure the reporting of injury to industrial workers; the Health Insurance Act and the Disablement Insurance Act

The European Coal and Steel Community has published lifting guidelines in *Force Limits in Manual Work*

Australia has several relevant advisory standards but no national legislation on the subject of MMH. Victoria has developed state legislation addressing MMH, *Regulations and Code of Practice for Manual Handling.* As such, the codes cannot be enforced outside of Victoria

Table 1.3 Limitations of Weight

Country	Conditions	Maximum weight (kg)
1. Adult male workers		
Brazil	For carrying up to 60 m	60
	For lifting by a worker on his own	40
Colombia	For carrying on the shoulders	50
	For lifting	25
Czechoslovakia (former)	Good gripping facilities for both hands	50
Ecuador	None	60
Federal Republic of Germany (former)	45+ years; occasional lifting/carrying	45
	45+ years; frequent lifting/carrying	25
	19–45 years; occasional lifting/carrying	55
	19–45 years; frequent lifting/carrying	30
German Democratic Republic (former)	Heavy/very heavy work; < 8% of shift	33
	Moderately heavy work; < 8% of shift	38
	Light/moderately heavy work; < 8% of shift	45
	Heavy/very heavy work; > 30% of shift	8
	Moderately heavy work; > 30% of shift	12
	Light/moderately heavy work; 30% of shift	15
Greece	For carrying meat of slaughtered animals	100
Honduras	None	50
Hungary	Carrying up to 90 m	50
Mozambique	None	55
Pakistan (Punjab state)	None	90
Philippines	In construction work (continuous lifting)	50
Poland	Continuous lifting/carrying up to 25 m	50
	On 60° slope (steps)/45° slope (ramp) to a maximum height of 4 m	30

Country	Conditions	Maximum weight (kg)
2. Adult female workers		
Austria	Nonexpectant mothers	15
Bulgaria	Over 16 years of age/even surface	20
Cameroon	Carrying, pushing, or pulling	25
Colombia	Over 18 years of age/lifting	12.5
	Over 18 years of age/carrying	15
Cote d'Ivoire	Carrying, pushing, or pulling	25
Czechoslovakia (former)	Lifting and carrying	15
	up to 90 cycles/shift	16
	up to 70 cycles/shift	18
	up to 44 cycles/shift	20
Federal Republic of Germany (former)	Occasional lifting	15
	Frequent lifting	10
German Democratic Republic (former)	Heavy/very heavy work; < 8% of shift	19
	Moderately heavy work; < 8% of shift	22
	Light/moderately heavy work; < 8% of shift	25
	Heavy/very heavy work; > 30% of shift	4
	Moderately heavy work; > 30% of shift	7
	Light/moderately heavy work; > 30% of shift	10
Hungary	Over 18 years of age, up to 60 m distance, and in pairs	40
Japan	18 years and older/continuous work	20
	18 years and older/intermittent work	30
Luxembourg	Expectant mothers/3 months after delivery	5
Malaysia	Over 18 years age/carrying	25
Mozambique	Over 21 years age/carrying	41.25
Pakistan (Punjab state)	17 years and older	23
Philippines	Lifting, carrying or moving	25
Poland	Lifting/carrying on flat surface	20
	Lifting/carrying on a slope	15
	Occasional lifting/carrying on flat surface	30
	Occasional lifting/carrying on a slope	25
USSR (former)	Lifting/carrying alternating with other operations	15
	Lifting over 1.5 m height	10
	Lifting/carrying continuously	10
3. Young working persons and children		
Bolivia	Boys under 16 years of age	10
	Girls under 16 years of age	5
	Women 16–20 years of age	10
Colombia	Intermittent lifting/16–18 years age	15
	Continuous lifting/16–18 years age	11.25
	Carrying/16–18 years of age	20
	Boys up to 16 years	15
	Girls up to 16 years	8
Cote d'Ivoire	Carrying/pushing/pulling/workers 16–18 years of age	20
	Boys 14–16 years of age	15

Country	Conditions	Maximum weight (kg)
Czechoslovakia (former)	Occasional carrying/workers up to 16 years of age	10
Ecuador	Boys 16–18 years of age	23
	Girls 18–21 years of age	12
	Boys under 16 years of age	16
	Girls under 18 years of age	9
Egypt	Carrying/boys 12–15 years of age	10
	Carrying/girls 12–15 years of age	7
Federal Republic of Germany (former)	Occasional/boys 15–18 years of age	35
	Frequent/boys 15–18 years of age	20
	Occasional/girls 15–18 years of age	15
	Frequent/girls 15–18 years of age	10
Finland	Boys 15–17 years of age	20
	Girls 15–17 years of age	15
Gabon	Carrying/boys 17–18 years of age	20
	Carrying/boys 16–17 years of age	15
	Carrying/girls 17–18 years of age	10
Greece	Lifting/carrying – workers 16–18 years of age	10
	Lifting/carrying – children up to 16 years of age	5
Hungary	Lifting/carrying up to 60 m – workers 16–18 years of age, alone	20
	Lifting/carrying up to 60 m in pairs – workers up to 16–18 years of age	40
	Lifting/carrying – boys 14–16 years of age, alone	15
	Lifting/carrying – boys 14–16 years of age, in pairs	30
	Lifting/carrying – girls 16–18 years of age, alone	15
	Lifting/carrying – girls 16–18 years of age, in pairs	30
	Carrying 1% slope – boys 14–16 years of age and girls 16–18 years of age	10
	Carrying 2% slope – boys 14–16 years of age and girls 16–18 years of age	5
Israel	Boys 16–18 years, < 2 hours/day	20
	Boys 16–18 years, > 2 hours/day	16
	Boys under 16 years, < 2 hours/day	12.5
	Boys under 16 years, > 2 hours/day	10
	Girls 16–18 years, < 2 hours/day	10
	Girls under 16 years, < 2 hours/day	8
Japan	Lifting/carrying, boys 16–18, intermittent	30
	Lifting/carrying, girls 16–18, intermittent	25
	Lifting/carrying, boys < 16, intermittent	15
	Lifting/carrying, girls < 16, intermittent	12
	Lifting/carrying, boys 16–18, continuous	20
	Lifting/carrying, girls 16–18, continuous	15
	Lifting/carrying, boys < 16, continuous	10
	Lifting/carrying, girls < 16, continuous	8
Malaysia	Carrying, boys 16–18 years of age	20
	Carrying, boys 14–16 years of age	15
	Carrying, girls 16–18 years of age	10
	Carrying, girls 14–16 years of age	8

Country	Conditions	Maximum weight (kg)
Mexico	Carrying, boys under 16 years of age	20
	Carrying, girls under 16 years of age	10
Pakistan	Lifting/carrying, boys 15–17 years of age	23
	Lifting/carrying, girls 15–17 years of age	18
	Lifting/carrying, children under 15	16
Poland	Carrying, workers 16–18 on slope < 30° and height < 5 m	8
	Carrying, boys under 16 on flat surface	16
	Carrying, girls under 16 on flat surface	10
	Carrying, boys under 16 on slope, steps or ramp	5
	Carrying, girls under 16 on slope, steps or ramp	3

(Source: ILO, 1988)

injuries and is intended for individual industries to produce more specific guidance to suit their own circumstances. The main points in the guide are summarized in a checklist for employers (Table 1.6).

The Council of the European Communities has also adopted a directive on the minimum health and safety requirements for the manual handling of loads where there is a risk particularly of back injury to workers (*Official Journal of the European Communities*, 1990). The directive obligates employers to: organize workstations in such a way as to make handling as safe and healthy as possible by considering factors listed in Table 1.7; ensure that workers and/or their representatives receive general indications and, where possible, precise information on

Table 1.4 Frequency multiplier

F lifts/min	Duration					
	< 1 h		1–2 h		2–8 h	
	$V < 76.2$ cm	$V \geq 76.2$ cm	$V < 76.2$ cm	$V \geq 76.2$ cm	$V < 76.2$ cm	$V \geq 76.2$ cm
≤ 0.2	1.00	1.00	0.95	0.95	0.85	0.85
0.5	0.97	0.97	0.92	0.92	0.81	0.81
1	0.94	0.94	0.88	0.88	0.75	0.75
2	0.91	0.91	0.84	0.84	0.65	0.65
3	0.88	0.88	0.79	0.79	0.55	0.55
4	0.84	0.84	0.72	0.72	0.45	0.45
5	0.80	0.80	0.60	0.60	0.35	0.35
6	0.75	0.75	0.50	0.50	0.27	0.27
7	0.70	0.70	0.42	0.42	0.22	0.22
8	0.60	0.60	0.35	0.35	0.18	0.18
9	0.52	0.52	0.30	0.30	0	0.15
10	0.45	0.45	0.26	0.26	0	0.13
11	0.41	0.41	0	0.23	0	0
12	0.37	0.37	0	0.21	0	0
13	0	0.34	0	0	0	0
14	0	0.31	0	0	0	0
15	0	0.28	0	0	0	0
> 15	0	0	0	0	0	0

Table 1.5 Coupling multiplier

Coupling type	*CM*	
	$V < 76.2$ cm	$V \geq 76.2$ cm
Good	1.00	1.00
Fair	0.95	1.00
Poor	0.90	0.90

weight of a load and the centre of gravity of the heaviest side when a package is eccentrically loaded; ensure that workers receive proper training and information on how to handle loads correctly and the risks they might be open to, particularly if these tasks are not performed correctly. The directive requires member states to bring into force the laws, regulations and administrative provisions needed to comply with the directive not later than 31 December 1992.

In Japan, there is an upper weight limit of lift prescribed by law (Hiraga, 1988). The limit is based on worker age, gender, and the task cycle (intermittent versus continuous work). The limits range from 8 kg for women under 16 years of age doing continuous work to 30 kg for men and women over 18 years of age.

There are two legislative acts in the Netherlands which have relevance to the MMH problem: the Health Insurance Act and the

Table 1.6 Checklist for employers

1. The task – does it involve:
   - holding load at a distance from trunk?
   - unsatisfactory bodily movement or posture, especially:
     - twisting the trunk?
     - stooping?
   - excessive movement of load, especially:
     - excessive lifting or lowering distances?
     - excessive carrying distances?
     - excessive pushing or pulling distances?
   - risk of sudden movement of load?
   - frequent or prolonged physical effort?
   - insufficient rest or recovery periods?
2. The load – is it:
   - heavy?
   - bulky or unwieldy?
   - difficult to grasp?
   - unstable, or with contents likely to shift?
   - sharp, hot or otherwise potentially damaging?
3. The working environment – are there:
   - space constraints preventing good posture?
   - uneven, slippery, or unstable floors?
   - variations in level of floors or work surfaces?
   - extremes of temperature, humidity, or air movement?
   - poor lighting conditions?
4. Individual capability – does the job:
   - require unusual strength, height, etc.
   - create a hazard to those who are pregnant or have a health problem?
   - require special knowledge or training for its safe performance?

(Source: Health and Safety Commission, 1988)

Table 1.7 Work, environment and personal factors to be considered in workstation organization

1. Characteristics of the load

   The manual handling of a load may present a risk particularly of back injury if it is:
   - too heavy or too large
   - unwieldy or difficult to grasp
   - unstable or has contents likely to shift
   - positioned in a manner requiring it to be held or manipulated at a distance from the trunk, or with a bending or twisting of the trunk
   - likely, because of its contents and/or consistency, to result in injury to workers, particularly in the event of a collision

2. Physical effort required

   A physical effort may present a risk particularly of back injury if it is:
   - too strenuous
   - only achieved by a twisting movement of the trunk
   - likely to result in a sudden movement of the load
   - made with the body in an unstable posture

3. Characteristics of the working environment

   The characteristics of the working environment may increase a risk particularly of back injury if:
   - there is not enough room, in particular vertically, to carry out the activity
   - the floor is uneven, thus presenting tripping hazards, or is slippery in relation to the worker's footwear
   - the place of work or the working environment prevents the handling of loads at a safe height or with good posture by the worker
   - there are variations in the level of the floor or the working surface, requiring the load to be manipulated on different levels
   - the floor or foot rest is unstable
   - the temperature, humidity, or ventilation is unsuitable

4. Requirements of the activity

   The activity may present a risk particularly of back injury if it entails one or more of the following requirements:
   - over-frequent or over-prolonged physical effort involving in particular the spine
   - an insufficient bodily rest or recovery period
   - excessive lifting, lowering, or carrying distances
   - a rate of work imposed by a process which cannot be altered by the worker

5. Individual risk factors

   The worker may be at risk if he/she:
   - is physically unsuited to carry out the task in question
   - is wearing unsuitable clothing, footwear, or other personal effects
   - does not have adequate or appropriate knowledge or training

(Source: *Official Journal of the European Communities*, 1990)

Disablement Insurance Act (Zuidema, 1985). These acts, like those of the UK, provide for the reporting of statistics on occupational injuries and accidents but provide no regulation of MMH activities *per se*.

Guidelines for manual lifting have also been developed by the European Coal and Steel Community (ECSC) setting force limits. The method for determining the force limit (FL) is set forth in the ECSC's pamphlet *Force Limits in Manual Work*. The FL is determined by a formula considering age, frequency of lifts, and the distances of the lifts.

$$FL = (A)(F)(50.15 + 0.332V' - 0.647H' - 0.0066V'^2 - 0.00372V'H' - 0.0000877V'^3 + 0.0000735V'^2H')$$

where FL = force limit (kg)
A = 1.0 for males < 40 years, 0.915 for males 41–50 years, and 0.782 for males 51–60 years
F = 1.0 for frequencies < 1 per minute and 0.7 for frequencies ≥ 1 per minute
H' = horizontal distance of load with respect to the shoulder (cm)
V' = vertical distance of the load with respect to the shoulder (cm)

The ECSC guide provides diagrams showing the FL for several different activities, including one- and two-handed lifts from standing, squatting, sitting, and kneeling positions.

Conclusions

The information presented here should by no means be considered a complete listing of statistics and legislation world-wide. The authors have attempted to be as comprehensive as possible with the information available. However, we feel certain there are other acts or legislations in progress or in existence. These were not left out by design. We simply were not aware of them or did not have access to translated documents.

Nonetheless, the information presented here provides a good indication of the efforts of the international community with regard to MMH activities. Growing recognition of the effects on workers engaged in these types of activities is only likely to improve efforts to collect information and take legislative actions.

References

Australian Council of Trade Unions, 1983. *Guidelines on Manual Handling*. Victoria: ACTU/VTHC Occupational Health and Safety Unit.

Biering-Sorensen, F., 1985. National statistics in Denmark – back trouble versus occupation. *Ergonomics*, **28**, 25–29.

Chaffin, D.B., 1987. Manual materials handling and the biomechanical basis for prevention of low-back pain in industry – an overview. *American Industrial Hygiene Association Journal*, **48**, 989–996.

David, G.C., 1985. UK national statistics on handling accidents and lumbar injuries at work. *Ergonomics*, **28**, 9–16.

Department of Labour, 1988. *Regulations and Code of Practice: Manual Handling*. Victoria: Department of Labour.

Health and Safety Commission, 1982. *Proposals for Health and Safety (Manual Handling of Loads) Regulations and Guidance*. London: HMSO.

Health and Safety Commission, 1988. *Handling Loads at Work* – Proposals for Regulations and Guidance. London: Health and Safety Executive.

Health and Safety Commission, 1991. *Handling Loads at Work* – Proposals for Regulations and Guidance. London: Health and Safety Executive.

Hettinger, T., 1985. Statistics on diseases in the Federal Republic of Germany with particular reference to diseases of the skeletal system. *Ergonomics*, **28**, 17–20.

Hiraga, M., 1988. Physical strain in carrying of heavy packages by human labor. *Proceedings of the International Ergonomics Association*, Sydney, Australia, pp. 239–241.

International Labour Office, 1988. *Maximum Weights in Load Lifting and Carrying*. Occupational Safety and Health Series No. 59. Geneva: International Labour Office.

Metzler, F., 1985. Epidemiology and statistics in Luxembourg. *Ergonomics*, **28**, 21–24.

National Board of Occupational Safety and Health, 1983. *Occupational Injuries in Sweden 1983*. Stockholm: National Board of Occupational Safety and Health.

National Institute for Occupational Safety and Health, 1981. *A Work Practices Guide for Manual Lifting*. Cincinnati: National Institute for Occupational Safety and Health.

National Institute for Occupational Safety and Health, 1996. *National Occupational Research Agenda (NORA)*. Cincinnati: National Institute for Occupational Safety and Health.

Official Journal of the European Communities, 1990. Council Directive, No. L 156, 21 June, pp. 9–13.

Rawling, R. and O'Halloran, P., 1988. Effective control of manual handling injury in the electricity industry. *Proceedings of the International Ergonomics Association*, Sydney, Australia, pp. 266–268.

Waters, T.R., Putz-Anderson, V., Garg, A., and Fine, L.J., 1993. Revised NIOSH equation for the design and evaluation of manual lifting tasks. *Ergonomics*, **36**, 749–776.

Zuidema, H., 1985. National statistics in the Netherlands. *Ergonomics*, **28**, 3–7.

Chapter 2

Factors to be considered in designing manual materials handling tasks

Introduction

Several worker-, work- and environment-related factors have been suggested by various researchers as risk factors in performing MMH activities. In this chapter, we review those factors that have been widely accepted by various agencies as the factors that must be controlled or modified in some systematic manner to reduce the risk of MMH injury. For a detailed discussion and research concerning all the factors that have been identified as possible risk factors, the reader is referred to the reference book *Manual Materials Handling* by Ayoub and Mital (1989).

Physique/anthropometry/strength

It is an intuitive expectation that well-built and strong individuals have a greater capability to perform MMH tasks than individuals who have a slight build and are relatively weak. While studies relating body size and type (obese, slim, and muscular) to capacity for performing MMH activities are non-existent, several specific body-size measures, such as shoulder height, chest width and depth, knee and knuckle heights, and abdominal depth, have been found useful in predicting an individual's capacity for manual lifting. In general, it has been observed that a proportionately built person (big chest and narrow waist) has greater lifting capacity than one who has a large torso and abdominal depth (Ayoub and Mital, 1989). Although individuals with a large torso and large abdominal depth tend to be stronger than other body types, they have a great deal of difficulty in repetitive load handling, and can get exhausted quickly, particularly when the load is located on the floor (Wyndham *et al.*, 1963; Kamon and Belding, 1971; Petrofsky and Lind, 1975; Garg, 1976). Heavier individuals certainly have an advantage over other body types in that these individuals are generally stronger and have enough mass to handle large objects and, therefore, have higher capacity for infrequently performed MMH tasks (Snook and Irvine, 1967; Troup and Chapman, 1969; Larsson *et al.*, 1979).

Several studies have also shown that, compared to shorter individuals, tall people are relatively weaker in lifting strength and more susceptible to back pain as they have to lean and reach further to pick up or set down a load (Switzer, 1962; Watson, 1977; Chaffin *et al.*, 1977; Merriam *et al.*, 1983). In fact, some earlier studies have shown that shorter weight lifters have an advantage (Keeney, 1955; Lietzke, 1956).

Many researchers have suggested that a person's ability to handle loads or exert forces is limited by his/her muscle strength, static (Ayoub and McDaniel, 1974; Chaffin *et al.*, 1977; Yates *et al.*, 1980; Mital and Manivasagan, 1983) or dynamic (Kamon *et al.*, 1982; Kroemer, 1985; Mital *et al.*, 1986). Weaker individuals or individuals who do not possess adequate static strength to perform manual lifting are also known to be prone to low-back pain (Chaffin, 1974).

The review of scientific literature indicates that taller, muscularly weak, or obese individuals are disadvantaged when performing materials handling jobs, particularly repetitive MMH jobs. Muscular built persons, on the other hand, have greater MMH capacity and are less prone to low-back pain.

Physical fitness/spinal mobility

It is widely believed that physical fitness improves a person's self esteem and may even lead to reduction in injuries and incapacity for work (Doelen and Wright, 1979). Physically fit individuals are also likely to report to work more regularly and might not report minor injuries as the injury-related pain tolerance level in such individuals is higher (Scott and Gibsbers, 1981). This may lead to a reduction in payroll costs (Cox *et al.*, 1981). Physically fit individuals also have a tendency to overestimate their physical capabilities and tend to take on tasks of greater risk and, consequently, suffer more severe injuries (Cady *et al.*, 1979). Least and moderately fit individuals, on the other hand, are most likely to have back or sciatic pain, particularly if they have had a history of back disorders (Lloyd and Troup, 1983). This finding, however, does not automatically lead to the conclusion that physically fit individuals (those engaged in sports and other physical activities) are less prone to back pain (Videman *et al.*, 1984). It also remains to be seen that activities specifically designed to enhance muscular strength or MMH capacity (Asfour *et al.*, 1984; Sharp and Legg, 1988) lead to a reduction in back disorders. For repetitive and endurance-oriented MMH activities, physical fitness means greater metabolic energy expenditure capability and greater preparedness for such activities. One could argue that this leads to the conclusion that physical fitness in such cases would reduce overexertion injuries. There is, however, no definitive scientific evidence to support this conclusion. **In general, a physically fit individual has a better outlook and may be better prepared to undertake muscular work than a less fit individual. From this standpoint alone, physical fitness should be emphasized.**

Lumbar mobility, though restricted in individuals with back pain, has been found to be more closely related to age than back pain (Wickstrom, 1978; Karvonen *et al.*, 1980). The likelihood of limited lumbar mobility also does not inspire a great deal of confidence in its value as a predictor of back pain as the probability that lumbar motion is restricted is about as high in individuals with no back pain as it is in individuals with a recent history of back pain (Horal, 1969; Troup *et al.*, 1981). **At the present time there is overwhelming evidence to suggest that limited lumbar mobility could, or does, lead to back pain or any other back disorder or poses a greater hazard to individuals engaged in carrying out MMH activities.**

Age/gender differences

The fact that people experience a decline in their capabilities as they age is well known. Aberg (1961) documented these reductions in capacity and observed that the capacity decreases after 20 years of age. The recognition that age has a profound effect on working capability has led to widely acceptable practices such as assigning the softest job to the oldest worker and restricting older workers from physically demanding jobs. Perhaps these practices are the reason for relatively fewer injuries among 50–60-year-old workers. Overloading workers in their early ages tends to accelerate the onset and rate of musculoskeletal injuries with ageing (Blow and Jackson, 1971; Brown, 1971). Protecting older workers from physically demanding jobs and assigning, and maybe overloading, younger and inexperienced workers to such jobs would be the logical explanation for the high rate of low-back pain incidence among 30–50-year-old workers (Brown, 1974; Rowe, 1983).

The effect of ageing on MMH capability is rather conflicting. While it is well established that ageing leads to reductions in physical work capacity, range of lumbar spinal motion, muscle strength, muscle contraction speed, shock absorbing characteristics of the lumbar disc, intra-abdominal pressure, load supporting capacity of the spine, and aerobic capacity (Ayoub and Mital, 1989), its effect on MMH capabilities does not appear to be significant. A number of studies reviewed by Ayoub and Mital (1989) show that between 18 and 61 years, age has no effect on manual lifting capacity of workers. Furthermore, the heart rates and oxygen uptakes at the maximum acceptable weights of lift also remain unaffected by age. It is very likely that older workers compensate for decline in working capability through improved skills and neuromuscular coordination. **In general, the effect of age on material handling capacity is very unclear. While the physical capabilities of individuals decline with age, this decline is not observed in the case of manual lifting capability. Given the fact that load bearing capacity of the spinal column declines with age, age should be treated as a risk factor and older workers, particularly those above the age of 50 years, should not be assigned to physically demanding jobs.**

Gender is the most critical worker attribute as it divides the working population into two distinct groups. Differences in the anthropometry, anatomy and physiology of men and women necessitate that women be treated differently from men. The fact that there will be some women who will be as strong as, or even stronger than, many men, and therefore quite capable of performing demanding physical jobs, should not be taken as evidence that both genders are equally at risk when performing demanding physical jobs.

Gender differences are also reflected in the MMH capabilities of men and women. In general, the MMH capability of women is substantially lower than that of men. Gender differences may be attributed primarily to differences in muscle strengths. On average, a woman's lifting strength is 60–76% of a man's lifting strength. Specific female strengths may be as low as 33% or as high as 86%. A similar relationship also exists for muscle power. Biomechanical linkage mechanism differences between males and females also contribute to differences between male and female MMH capacity. Furthermore, for the same task, women work closer to their aerobic capacities than men and, therefore, are at a greater risk (Ayoub and Mital, 1989). The working capacity of females also falls during pregnancy due to increases in basal metabolism and in abdominal and pectoral girths (Troup and Edwards, 1985).

Given the profound differences in the working and MMH capacities of men and women, it is essential that special attention be given to accommodating women in jobs that are performed by both men and women. A broad discrimination, on the basis of gender, is neither recommended nor required.

Psychophysical factors/motivation

The effects of psychological factors on psychophysical MMH capacity and maximal muscular performance are broad but not clearly understood. Table 2.1 summarizes the effects of some of these factors.

Table 2.1 Psychological factors affecting maximum muscular strength (Kroemer and Marras, 1981)

Factor	Likely effect
Feedback of results	Positive
Instructions on how to exert strength	Positive
Arousal of ego involvement, aspiration	Positive
Pharmaceutical agents (drugs)	Positive
Startling noise, subject's outcry	Positive
Hypnosis	Positive
Setting of goals, incentives	Positive or negative
Competition, contest	Positive or negative
Verbal encouragement	Positive or negative
Spectators	?
Deception by researcher	?
Fear of injury	?
Deception by subject	Negative

Some empirical evidence supporting the conclusions in Table 2.1 has been summarized by Ayoub and Mital (1989). Bakken (1983) made informal observations concerning the attitude of the individual and its effect on MMH capacity. He observed that attitude was a superior singular predictor of maximum acceptable weight of load than use of any other singular measurement including strength, anthropometric measures, climatic, or task descriptors. Unfortunately, no analyses were conducted to determine the 'content' of attitudes, nor was the attitude questionnaire formally validated or tested.

Feyer *et al.* (1992) attempted to associate nonphysical factors with low-back pain in white-collar and blue-collar workers. They determined that worker dissatisfaction was not related to the presence of low-back pain.

In the context of MMH activities, there are no direct and quantitative relationships between the factors listed in Table 2.1 and MMH capacity. There is also no specific information on the extent to which these factors influence MMH capacity. Many of the factors listed in Table 2.1 that have a positive effect, such as startling noise and use of drugs, are clearly the kind of factors that must be avoided in the workplace. Factors, such as competition and contest, also must be avoided as they may encourage individuals to overload themselves. Factors, such as incentives and setting the goals, should be introduced provided they have a positive effect on the attitude and do not encourage individuals to overexert themselves.

Training and selection

The overall effects of physical training are positive and lead to enhancement in muscular strength, MMH capacity, cardiovascular capability, and endurance time (Sharp and Legg, 1988; Ayoub and Mital, 1989; Genaidy, 1990). There is, however, less than overwhelming evidence to conclude that enhanced working capacity from physical training also leads to a reduction in back disorders and improvement in productivity (Snook *et al.*, 1978). Only a handful of investigations have reported a reduction in low-back pain following physical training (Karvonen *et al.*, 1977; Dehlin *et al.*, 1978; Meyers *et al.*, 1981).

However, since training has an educational value and enhances cardiovascular and muscular capabilities, MMH activities are perceived to become easier (reduced physical stress) and require less effort with training. Physical training, therefore, is highly desirable. The training programme should include not only physical training but training in safe handling techniques and use of materials handling aids as well and should be extended not only to new employees but also to existing workers. Classroom instruction on the hazards of MMH activities should be an integral part of a training programme.

In general, if a job requires strong or very strong people for satisfactory completion, it must be redesigned as the risk of injury in such cases is great. However, often it is impossible or impractical to design or redesign a job. In such cases, or in cases when efforts must be undertaken to prevent assignment of an individual with a history of back disorders to an MMH job, selection and screening of individuals is the only recourse.

The primary goals of screening and selection are: (1) to ensure that a person with a history of relevant disorder, disease or incapacity is not assigned to any MMH activity and (2) people who are assigned to an MMH job have the necessary strength required by the job. The various screening procedures have been reviewed in detail by Ayoub and Mital (1989) and the reader is referred to this publication for details. It should be realized that even a complete physical examination accompanied by lumbosacral radiographic evaluation predicts only 10% of low-back injury sufferers (Rowe, 1969). Strength testing, a very pervasive screening tool, also does not provide a direct relationship between a person's capability and the risk of back disorder. Furthermore, strength testing excludes important measures of a person's working capabilities, such as cardiovascular capacity.

The above discussion should not lead to the conclusion that strength testing, as a screening tool, has no utility. A number of researchers have shown that muscular strength of a person is an important predictor of his/her MMH capability. However, in recent years, emphasis has shifted from isometric (static) strength testing to dynamic (isokinetic and isoinertial) strength testing as dynamic strengths are more accurate predictors of MMH capabilities. Specifically, dynamic strengths that simulate the job (simulated job dynamic strengths) are more relevant in estimating a person's MMH capability than either physical attributes or isometric lifting strength; lifting strength may be important for a lifting task, but pulling strength would be important for pulling tasks (Ayoub and Mital, 1989).

In general, existing screening and selection devices, many of which are widely in use, are inadequate. Furthermore, these devices and techniques have shown no close correlation with injury occurrences. Of the various screening tools, simulated job dynamic strengths appear to provide a more accurate estimate of a person's MMH capacity. Procedures, such as X-rays, may be potentially more dangerous to the individual than the information they provide. Certainly these kind of procedures are necessary if there is evidence of spinal disorder or disease.

Effects of static work

Almost all MMH activities contain both a static and a dynamic component. In some tasks, the dynamic component is the major one (e.g. in repetitive lifting tasks), while in other tasks, the static component is the dominating component (e.g. in holding tasks). As evident from these two examples; static work or effort is characterized by contraction of muscles over extended periods, such as when a postural stance is adopted for a prolonged period. The physiological effects of static work are well known and include compression of blood vessels, lack of oxygen supply to muscle cells, accelerating loss of strength, and, eventually, pain. The lack of oxygen and buildup of waste products during static work make it increasingly strenuous.

In the context of MMH activities, very limited information is available on the effects of static work on MMH capacity. A few studies have shown that static endurance (time a load can be held without significant movement) is an important predictor of maximum acceptable weights of lift (Mital and Ayoub, 1980; Mital and Manivasagan, 1984). Ayoub *et al.* (1987) found that holding time decreases significantly as the load being held becomes heavier. Furthermore, people have difficulty in holding the object in place as it gets heavier. This finding is significant as in many instances workers are required to hold an object in place while loading it on to a machine or fastening it to another object/surface. Sustained sedentary work in a non-neutral trunk posture (non-erect) has also been reported to be associated with low-back pain (Burdorf *et al.*, 1993).

In general, static work should be avoided as much as possible. If it is not possible to avoid it, mechanical aids should be used. Carrying and holding are two specific MMH activities that should be aided by mechanical equipment.

Posture/handling techniques

Body postures not only change force requirements, they cause the work to become very strenuous, particularly when the work is predominantly static in nature (Burdorf *et al.*, 1993). For MMH activities that have a dominating dynamic component, the body may assume different postures. For instance, manual lifting can be performed in squat, stooped or free-style (semi-squat) posture. The force requirements as well as metabolic energy requirements for all these postures are very different (Ayoub and Mital, 1989). For instance, the squat posture minimizes stress on the spinal column and the back muscles, it imposes a metabolic energy penalty on the body (a part of the body is also lifted along with the load. In the squat posture, the part of the body that is lifted weighs significantly more than when the stooped posture is employed.) The stooped posture, on the other hand, requires less metabolic energy, but leads to severe stress on the back. The physiological costs are also affected when the posture is non-erect; they are higher in non-erect postures than in erect postures. Turning and twisting while handling materials also leads to an increase in spinal stresses and intra-abdominal pressures (IAP). The task is also perceived to be more difficult (Ayoub and Mital, 1989). Turning, twisting, and bending are also associated with increased incidence of low-back disorders (pain, ache, and discomfort; Christensen *et al.*, 1995).

The exception is the seated posture. Generally, the IAP is higher in the seated posture as compared to the standing posture. However, back supports can reduce IAP (Boudrifa and Davis, 1984). As a rough rule, force requirements increase as the load moves farther away from the body. Keeping the load between the knees would help minimize these forces.

The stoop posture is advantageous when the load must be lifted repeatedly. The squat posture is desirable when the load can be fitted between the knees and is handled only occasionally. Loads that cannot be fitted between the knees and must be lifted repetitively should be handled by two individuals or must be moved with the help of mechanical equipment. In general, an extreme range of movements (turning/twisting, jerky motions) and fixed postures should be avoided. As much as possible, moving loads from the floor should be avoided. If possible, load movement should be restricted between knee and shoulder height. Lifting loads to reach and overreach heights, particularly when the load is to be lifted from the floor, must be avoided. Pushing force should be exerted in near erect posture if possible, with handles located at a height of approximately 1 m. Pulling loads should be avoided (Ayoub and Mital, 1989).

If a load is awkward in shape and size or is too heavy for one person to lift, mechanical aids should be used. If for some reason mechanical aids cannot be used, for instance when the work is performed in narrow and confined spaces, there is no option other than assigning two or more individuals to handle it. **In such cases, the load should be lifted using the squat posture (Bendix and Eid, 1983), the weight of the load should be less than the sum of the capacities of the individuals, and the people engaged in lifting it should be similar in height. Coordination of the activity through counting or some sort of verbal signalling is also highly desirable.**

Loading characteristics

A load is generally characterized by its shape, size and weight. The weight of the load is perhaps the most important load characteristic in most situations in which the load will be handled manually. This attribute, to a large extent, also determines how strenuous a job is. Part II of this guide is devoted to weight and force limits that should not be exceeded if a material handling task is to be performed manually. In this section, the discussion is limited to the shape and size of the load.

Limited information is available regarding the influence of load shape on MMH capacity and how it is correlated to injury. A few investigators have reported that shape of the object influences MMH capacity as well as the physiological costs. The findings indicate that collapsible containers (bags) allow greater MMH capacity but at higher physiological costs as compared to non-collapsible/rigid (box) containers. **The limited number of investigations that have studied load shape do not permit a general conclusion to be drawn. However, considerations of good grip, smooth movements while handling the load, load stability during handling, etc. dictate that the load should be rigid and symmetrical in shape. Symmetrical loads also permit greater load stability, provided the load is distributed uniformly. If the load is not distributed uniformly, the heavier end should be closer to the body. If the load centre of gravity is offset in the frontal plane, the heavier end should be held by the stronger arm.**

As in the case of load shape, only a handful of studies have looked at the size of the load. It is universally agreed that the load should be as small as possible. Larger loads not only make the handling awkward, they lead to higher physiological costs and greater spinal stresses

(Ayoub and Mital, 1989). If a load cannot be fitted between the knees, it is probably too large. The load dimension in the sagittal plane should be as small as possible to minimize stress on the back muscles. Muscular strength also decreases as the distance in the sagittal plane increases.

Specifically, the load dimension in the sagittal plane should not exceed 50 cm (Mital and Ayoub, 1981; Ostrom *et al.*, 1991). The load dimension between the hands (in the frontal plane) is less critical, but nevertheless should be minimized. The height of the load is the least critical of the three dimensions and practical considerations, such as anthropometry and ability to clearly view obstructions in the path, should determine this dimension.

Handles/coupling

Good handles or couplings are essential to provide load and postural stability during materials handling (Mital and Ayoub, 1981). The literature consistently supports the provision of handles on containers. A good coupling between the individual and the floor is also essential if the load is to be carried, pushed or pulled.

Handles can have a significant impact on MMH capacity. According to Garg and Saxena (1980), provision of handles can increase lifting capacity anywhere from 4% to 11.5%. Bakken (1983) observed a reduction of 15% in the lifting capacity for containers without handles. Nearly an 11% reduction in capacity has also been reported by Smith and Jiang (1984). Snook and Ciriello (1991) reported a range of reduction of 4% to 30% in the maximum acceptable weight of lift when boxes without handles were lifted. In terms of spinal stresses and metabolic cost, handles are desirable, but make little difference on either response.

Given that handles should be an essential feature of containers, the question is what should be their size and where on the container should they be located. Drury *et al.* (1985) reported an extensive study on handles and recommended that **cut-out handles should be**

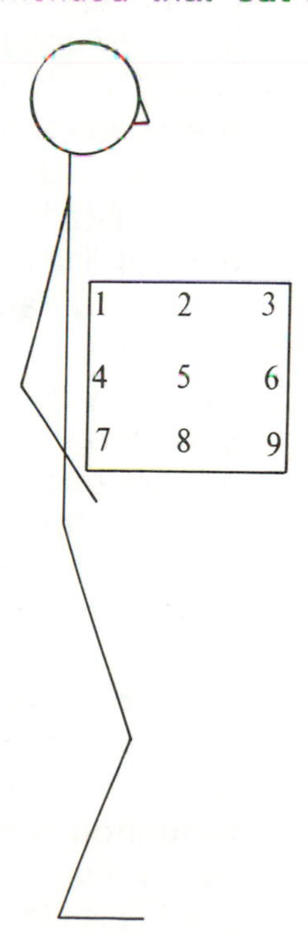

Figure 2.1 Handle position (Drury et al., *1982).*

115 mm long and 25–38 mm wide (or in diameter, in case cylindrical handles can be provided; cylindrical handles should also have a 30–50 mm clearance all round) and should have a pivot angle of 70° from the horizontal axis of the box. The handles should be located in the 3/8 position on both sides (Figure 2.1) to provide both vertical and horizontal stability for the load. A reduction of up to 15% in lifting capability may result if the containers do not have handles or firm holds to initiate the lift.

In order to prevent slipping while carrying, pushing or pulling, the coefficient of friction between the shoe sole and the floor should be at least 0.3. **Preferably, the coefficient of friction should be at least 0.5.** The shoe and floor material that can provide the necessary coefficient of friction would vary depending upon whether the floor is covered with mud, oil or other liquids. **In general, hardened rubber, dense vinyl plastic or leather shoes provide a good coupling** (Ayoub and Mital, 1989).

Repetitive handling

Frequency of handling is perhaps the most critical task characteristic that influences an individual's capability to perform MMH activities. Ayoub and Mital (1989) provide an extensive review of the effects of frequency on fatigue and MMH capabilities. Control of frequency is essential if the goal is to reduce fatigue and overall job strain. Part II of this guide provides MMH job design data as a function of frequency. **In general, the MMH capacity increases as the frequency of handling decreases. Furthermore, a reduction in frequency of handling also leads to increase in endurance time (Jomoah *et al.*, 1991). MMH activities that require frequent handling should either be redesigned to reduce the frequency or mechanical equipment should be used to aid the handling.**

Asymmetrical lifting/load asymmetry

Lifting or carrying objects asymmetrically is the rule rather than an exception in industrial settings. Asymmetrical materials handling leads to reduced lifting and carrying capabilities and isometric strength, increased intra-abdominal and intra-discal (shear) pressures, and increased electromyographic activity of erector spinae and external obliques. The metabolic energy expenditure, however, is generally not affected (Ayoub and Mital, 1989). As far as the decline in manual lifting capabilities is concerned, it could range from 8.5% (Mital and Fard, 1986) to 22% (Garg and Badger, 1986) subject to movement of the feet. If the feet are allowed to move, the decline in capability is far less than if the feet remain fixed. **In a realistic situation, the starting and ending locations of the load change continuously. The feet, therefore, move constantly, and it is very unlikely that a person will continue to handle material without moving the feet at all. In the event that the feet remain fixed, the spinal column twists and it is this twist that leads to a greater reduction in material handling capability. The materials handler, therefore, should be advised not to keep the feet in a locked position. If the feet move, not only is the reduction in MMH capability smaller, but the task is also less stressful. The reduction in manual lifting capability in such cases is expected to be no more than 15% for a 90° turning.**

It is also important to realize that most loads are not symmetrical. While most research has dealt with and accounted for a centre of gravity (c.g.) movement in the sagittal plane (away from the body), very little attention has been paid to c.g. movement in the frontal plane (sideways). **Sideways movement not only puts greater burden on the hand closer to the c.g., but it also subjects the spine to a lateral bending moment. This lateral bending moment results in a decline in MMH capability of up to 16% (Ayoub and Mital, 1989). To reduce physical stress on the hands and arms, the stronger hand should be closer to the load c.g.**

Confined environments/ spatial restraints

Performing MMH activities with some form of spatial restraint is a common occurrence in industry. In spite of our awareness of spatial restraints in the workplace, only limited attention has been directed at quantifying the effects of spatial restrictions on MMH task performance. Ridd (1985) reported that limited head room has a negative effect on acceptable lifting capacity. He later recommended a reduction of 1% in weight of the load for each degree of flexion imposed by the limited headroom (Ridd, 1991). Mital (1986) also found that when loads are carried through a narrow and confined passage, the acceptable load carrying capacity declines substantially. Mital and Wang (1989) concluded that shelf opening clearance has almost the same impact on lifting capacity as the absence of handles on the load. They observed that acceptable lifting capacity declined by almost 13% when the shelf opening clearance was 3 mm, as compared to unrestricted shelf opening.

For inserting boxes by hands in openings, a variety of recommendations exist. Most of these recommendations are contradictory and vary from 45 mm to 115 mm. **The shelf opening clearance for inserting boxes by hands should be approximately 30 mm. If the workplace layout does not allow erect posture, for example, due to limited headroom, the load should be reduced by 1% for each degree of trunk flexion from the erect posture.**

Safety aspects

Load handling may involve a number of hazards. Troup and Edwards (1985) have outlined several of them. The following is a modified list of hazards:

(a) surfaces likely to abrade, puncture or cut
(b) surfaces likely to be exposed to gusts of wind (e.g. sheets of glass, boarding, etc.)
(c) surface features likely to catch in clothing
(d) lack of adequate handles or gripping surface
(e) liability to swing (when a long load is held centrally; e.g. ladders)
(f) obstruction of vision (e.g. when climbing stairs, steps or going down a slope)
(g) contents of the container (toxins, acid, etc.)
(h) disabling glare that might cause collision
(i) slippery floors
(j) high temperatures, open flames, etc.
(k) atmospheric pollutants (e.g. dust, smoke, fumes, etc.)
(l) unstable platforms
(m) high noise level and vibration (may reduce vigilance, affect grip strength, etc.)

Routine safety analysis should be carried out to eliminate or minimize these hazards.

Protective equipment

Protective equipment includes shoes, gloves, vest and trousers, goggles, respirators, aprons and overalls and masks. Each of these devices serves a specific need. Shoes, for example, provide proper coupling between the individual and the floor. Similarly, goggles protect the eyes from disabling glare and masks and respirators protect the individual from inhaling toxic fumes, smoke, dust, etc. **Depending upon the hazard (see the previous section on safety aspects), proper protective equipment should be selected. Protective clothing should permit free movement, gloves should fit properly and allow maintenance of dexterity, shoes should be of the non-slip type, comfortable and waterproof. Protective clothing should be easily removable and, if possible, allow for personal cooling (protection from body metabolic heat buildup).**

Handling in a hot environment

It is well known that heat load (a combination of air temperature, air velocity, radiant heat and relative humidity) influences a person's physiological and psychological behaviour. At the very least, heat stress causes discomfort which leads to reduced work-rate, increased irritability, carelessness, a feeling of fatigue and increased accident rates (Ayoub and Mital, 1989). MMH activities that are performed in hot and temperate climates lead to elevated heart rates and increased rectal temperatures. MMH capacity also declines. According to Snook and Ciriello (1974), lifting capacity may decline by 20%, pushing capability by 16%, and carrying capability by 11% when the temperature increases from 17.2 to 27°C. Snook and Ciriello's subjects were, however, unacclimatized. Working with well-acclimatized individuals, Hafez (1984) determined that reduction in capacity is not as great as reported by Snook and Ciriello (1974). Hafez observed a 12% reduction in manual lifting capabilities of well-acclimatized subjects when the heat stress increased to 32°C wet bulb globe temperature (WBGT); no decline in capacity was observed up to 27°C WBGT.

Adequate rest and replenishment of body fluids are essential when work is to be performed in hot climate. The working time is the time it takes the core temperature to rise to 38.5°C; the resting time is the time it takes the core temperature to return to normal (37°C) while the worker is resting in an area, removed from the work area, which has comfort zone conditions. Cooling jackets may also be used to keep the core temperature from rising.

Task duration

The duration of task performance is an important consideration in designing an MMH job. As the task duration increases, the metabolic energy expenditure level of the operator also increases. This increase is primarily due to the accumulative effects of fatigue. Furthermore, the metabolic energy expenditure level that can be maintained decreases with the task duration (Ayoub and Mital, 1989). **The task burden (demand), therefore, should be reduced as the task duration increases.** It has been observed that when workers are given a choice to adjust the weight of the load, the weight declines with the task duration. **The reduction in weight is to maintain constant levels of fatigue and is as much as 3.4% per hour for males and 2% per hour for females. If the weight is not reduced, the metabolic energy expenditure rate and heart rate will increase** (Ayoub and Mital, 1989).

Conversely, if the task duration decreases, the weight of the

load and workload can increase. The increase in weight, however, must be within the biomechanical tolerance limits of the musculoskeletal system.

Work organization

Work organization has the following main considerations: workplace geometry, fixed postures, rest pauses, and job rotation. The effect of inadequate space on MMH capacity was discussed earlier in this chapter. Constant reorientation of load, postural instability, slower and cautious movements, increased IAP and reduced MMH capacity are some of the outcomes of inappropriate workplace geometry (inadequate space). Loading, unloading and maintenance are some of the routine activities that are performed in inadequate spaces. Even though workplace geometry has not been the subject of any scientific study, some common sense recommendations can be made. **Specifically, educating employees in safe procedures and reducing job demands (weight, frequency, reach requirements, rotation and asymmetry) are essential to reduce the hazards of MMH activities that are performed in workplaces that do not have adequate space. Allowing enough room to manoeuvre in the workplace and providing enough space for materials, shelves, tables, etc., is another prime requirement.**

Fixed postures should be avoided (see earlier section in this chapter) and the MMH activity should be redesigned to minimize static work component; otherwise, use of mechanical aids should be considered.

Adequate rest allowances should be provided to overcome the effect of fatigue. The procedure outlined by Ayoub and Mital (1989) may be used for this purpose. Job rotation should also be considered to minimize monotony, inattentiveness, and fatigue on a specific group of muscles.

References

Aberg, U., 1961. Physiological and mechanical studies of material handling. *Ergonomics*, **4**, 282.

Asfour, S.S., Ayoub, M.M. and Mital, A., 1984. Effects of an endurance and strength training programme on the lifting capability of males. *Ergonomics*, **27**, 435–442.

Ayoub, M.M. and McDaniel, J.W., 1974. Effects of operator stance on pushing and pulling tasks. *Transactions of the American Institute for Industrial Engineers*, **6**, 185–195.

Ayoub, M.M. and Mital, A., 1989. *Manual Materials Handling*. London: Taylor & Francis.

Ayoub, M.M., Selan, J.L. and Jiang, B.C., 1987. Manual materials handling. In *Handbook of Human Factors*, edited by G. Salvendy, pp. 790–818. New York: John Wiley.

Bakken, G.M., 1983. Lifting capacity determination as a function of task variables. PhD dissertation, Texas Tech University, Lubbock, Texas.

Bendix, T. and Eid, S.E., 1983. The distance between the load and the body with three bi-manual lifting techniques. *Applied Ergonomics*, **14**, 185–192.

Blow, R.J. and Jackson, J.M., 1971. Rehabilitation of registered dock workers. *Proceedings of the Royal Society of Medicine*, **64**, 753–760.

Boudrifa, H. and Davis, B.T., 1984. The effect of backrest inclination, lumbar support and thoracic support of the intra-abdominal pressure while lifting. *Ergonomics*, **27**, 379–387.

Brown, J.R., 1971. *Lifting As An Industrial Hazard*. Ontario, Canada: Labour Safety Council.

Brown, J.R., 1974. Lifting as an industrial hazard. *American Industrial Hygiene Association Journal*,

Burdorf, A., Naaktgeboren, B., and de Groot, H.C.W.M. 1993. Occupational risk factors for low-back pain among sedentary workers. *Journal of Occupational Medicine*, **12**, 1213–1220.

Cady, L.D., Biscoff, D.P., O'Connell, E.R., Thomas, P.C. and Allen, J.H., 1979. Strength and fitness and subsequent back injuries in firefighters. *Journal of Occupational Medicine*, **21**, 269–272.

Chaffin, D.B., 1974. Human strength capability and low-back pain. *Journal of Occupational Medicine*, **16**, 248–254.

Chaffin, D.B., Herrin, G.D., Keyserling, W.M. and Foulke, J.A., 1977. *Pre-employment Strength Testing in Selecting Workers for Materials Handling Jobs*. Cincinnati: National Institute for Occupational Safety and Health. Publication CDC-99-74-62.

Christensen, H., Pedersen, M.B., and Sjøgaard, G., 1995. A national cross-sectional study in the Danish wood and furniture industry on working postures and manual materials handling. *Ergonomics*, **38**, 793–805.

Cox, M., Shephard, R.J. and Storey, P., 1981. Influence of an employee fitness program upon fitness, productivity, and absenteeism. *Ergonomics*, **24**, 795–806.

Dehlin, O., Berg, S., Hedenrud, B., Anderson, G. and Grimby, G., 1978. Muscle training, psychological perception of work and low back symptoms in nursing aids. *Scandinavian Journal of Rehabilitation Medicine*, **10**, 201–209.

Doelen, J.V. and Wright, G.R., 1979. *Fitness and Occupational Injuries: A Review*. Safety Studies Service Report. Ontario, Canada: Ministry of Labour.

Drury, C.G., Law, C.H. and Pawenski, C.S., 1982. A survey of industrial box handling. *Human Factors*, **24**, 553–565.

Drury, C.G., Begbie, K., Ulate, C. and Deeb, J.M., 1985. Experiments on wrist deviation in manual materials handling. *Ergonomics*, **28**, 577–589.

Feyer, A.M., Williamson, A., Mandryk, J., de Silva, I., and Healy, S., 1992. Role of psychosocial risk factors in work-related low-back pain. *Scandinavian Journal of Work, Environment, and Health*, **18**, 368–375.

Garg, A., 1976. A metabolic rate prediction model for manual materials handling jobs. PhD dissertation, University of Michigan, University Microfilms.

Garg, A. and Badger, D., 1986. Maximum acceptable weights and maximum voluntary strength for asymmetric lifting. *Ergonomics*, **29**, 878–892.

Garg, A. and Saxena, U., 1980. Container characteristics and maximum acceptable weight of lift. *Human Factors*, **22**, 487–495.

Genaidy, A.M., 1990. A training programme to improve human physical capability for manual handling jobs. *Ergonomics*, **34**, 1–11.

Hafez, H.A., 1984. Manual lifting under hot environmental conditions. PhD dissertation, Texas Tech University, Lubbock, Texas.

Horal, J., 1969. The clinical appearance of low back disorders in the City of Gothenburg, Sweden. *Acta Orthopedicae Scandinavica*, (Suppl) 118.

Jomoah, I.M., Asfour, S.S. and Genaidy, A.M., 1991. Physiological models and guidelines for the design of high frequency arm lifting tasks. In *Advances in Industrial Ergonomics and Safety* III, edited by W. Karwowski and J.W. Yates, pp. 309–315. London: Taylor & Francis.

Kamon, E. and Belding, H.S., 1971. The physiological cost of carrying loads in temperate and hot environment. *Human Factors*, **13**, 153–161.

Kamon, E., Kiser, D. and Pytel, J., 1982. Dynamic and static lifting capacity and muscular strength of steelmill workers. *American Industrial Hygiene Association Journal*, **43**, 853–857.

Karvonen, M.J., Jarvinen, T. and Nummi, J., 1977. Follow-up study on the back problems of nurses. *Instructional Occupational Health*, **14**, 8.

Karvonen, M.J., Viitasalo, J.T., Komi, P.V., Nummi, J. and Jarvinen, T., 1980. Back and leg complaints in relation to muscle strength in young men. *Scandinavian Journal of Rehabilitation Medicine*, **12**, 53–60.

Keeney, C.E., 1955. Relationship of body weight to strength–body weight ratio in championship weightlifters. *Research Quarterly*, **26**, 54–59.

Kroemer, K.H.E., 1985. Testing individual capability to lift material: repeatability of a dynamic test compared with static testing. *Journal of Safety Research*, **16**, 1–7.

Kroemer, K.H.E. and Marras, W.S., 1981. Evaluation of maximal and submaximal static muscle exertions. *Human Factors*, **23**, 643–653.

Larsson, K., Grimby, G. and Karlsson, J., 1979. Muscle strength and speed of movement in relation to age and muscle morphology. *Journal of Applied Physiology*, **46**, 451–456.

Lietzke, M.H., 1956. Relation between weight-lifting totals and body weight. *Science*, **124**, 486–487.

Lloyd, D.C.E.F. and Troup, J.D.G., 1983. Recurrent back pain and its prediction. *Journal of Social and Occupational Medicine*, **33**, 66–74.

Merriam, W.F., Burwell, R.G., Mulholland, R.C., Pearson, J.C.G. and Webb, J.K., 1983. A study revealing a tall pelvis in subjects with low back pain. *Journal of Bone and Joint Surgery*, **65-B**, 153–156.

Meyers, J., Riordan, R., Mattmiller, B., Belcher, O., Levenson, B.S. and White, A.W., 1981. Low back injury prevention at Southern Pacific Railroad - 5 year experience with a Back School Model. Paper presented at the International Society for the Study of the Lumbar Spine Annual Meeting, Paris, May 16–20.

Mital, A., 1986. Subjective estimates of load carriage in confined and open spaces. In *Trends in Ergonomics/Human Factors III*, edited by W. Karwowski, pp. 827–833. Amsterdam: North-Holland.

Mital, A. and Ayoub, M.M., 1980. Modeling of isometric strength and lifting capacity. *Human Factors*, **22**, 285–290.

Mital, A. and Ayoub, M.M., 1981. Effect of task variables and their interactions in lifting and lowering loads. *American Industrial Hygiene Association Journal*, **42**, 134–142.

Mital, A. and Fard, H.F., 1986. Psychophysical and physiological responses to lifting symmetrical loads symmetrically and asymmetrically. *Ergonomics*, **29**, 1263–1272.

Mital, A. and Manivasagan, I., 1983. Maximum acceptable weight of lift as a function of material density, center of gravity location, hand preference, and frequency. *Human Factors*, **25**, 33–42.

Mital, A. and Manivasagan, I., 1984. Development of non-linear polynomials in identifying isometric strength behaviour. *Computers and Industrial Engineering – An International Journal*, **8**, 1–9.

Mital, A. and Wang, L.W., 1989. Effects on load handling of restricted and unrestricted shelf opening clearances. *Ergonomics*, **32**, 39–49.

Mital, A., Channaveeraiah, C., Fard, H.F. and Khaledi, H., 1986. Reliability of repetitive dynamic strengths as a screening tool for manual lifting tasks. *Clinical Biomechanics*, **1**, 125–129.

Ostrom, L.T., Smith, J.L. and Ayoub, M.M., 1991. The effect of box height on maximum acceptable weight of lift. In *Advances in Industrial Ergonomics and Safety III*, edited by W. Karwowski and J.W. Yates, pp. 263–267. London: Taylor & Francis.

Petrofsky, J.S. and Lind, A.R., 1975. Aging, isometric strength and endurance, and cardiovascular responses to static effort. *Journal of Applied Physiology*, **38**, 91–95.

Ridd, J.E., 1985. Spatial restraints and intra-abdominal

pressure. *Ergonomics*, **28**, 149–166.

Ridd, J.E., 1991. Physical work capacity in stooped and asymmetric postures. In *Designing for Everyone*, edited by Y. Queinnec and F. Daniellou, pp. 81–83. London: Taylor & Francis.

Rowe, M.L., 1969. Low back pain in industry: a position paper. *Journal of Occupational Medicine*, **11**, 161–169.

Rowe, M.L., 1983. *Backache at Work*. Fairport, New York: Perinton Press.

Scott, V. and Gibsbers, K., 1981. Pain perception in competitive swimmers. *British Medical Journal*, **283**, 91–93.

Sharp, M. and Legg, S., 1988. Effects of psychophysical lifting training on maximum repetitive lifting capacity. *American Industrial Hygiene Association Journal*, **49**, 639–644.

Smith, J.L. and Jiang, B.C., 1984. A manual materials handling study of bag lifting. *American Industrial Hygiene Association Journal*, **45**, 505–508.

Snook, S.H. and Ciriello, V.M., 1974. The effects of heat stress on manual handling tasks. *American Industrial Hygiene Association Journal*, **35**, 681–685.

Snook, S.H. and Ciriello, V.M., 1991. The design of manual handling tasks: revised tables of maximum acceptable weights and forces. *Ergonomics*, **34**, 1197–1213.

Snook, S.H. and Irvine, C.H., 1967. Maximum acceptable weight of lift. *American Industrial Hygiene Association Journal*, **28**, 322–329.

Snook, S.H., Campnelli, R.A. and Hart, J.W., 1978. A study of three preventive approaches to low back injury. *Journal of Occupational Medicine*, **20**, 478–481.

Switzer, S.A., 1962. *Weight Lifting Capacities of a Selected Sample of Human Males*, Report No. AD-284054. Wright Patterson Air Force Base, Ohio: Aerospace Medical Research Laboratory.

Troup, J.D.G. and Chapman, A.E., 1969. The static strength of the lumbar erectores spinae. *Journal of Anatomy*, **105**, 186.

Troup, J.D.G. and Edwards, F.C., 1985. *Manual Handling and Lifting: An Information and Literature Review With Special Reference to the Back*. London: HMSO.

Troup, J.D.G., Martin, J.W. and Lloyd, D.C.E.F., 1981. Back pain in industry: a prospective survey. *Spine*, **6**, 61–69.

Videman, T., Nurminen, T., Tola, S., Kuorinka, I., Vanharanta, H. and Troup, J.D.G., 1984. Low back pain in nurses and some loading factors of work. *Spine*, **9**, 400–404.

Watson, A.W.S., 1977. The relationship of muscular strength to body size and somatype in post-puberal males. *Irish Journal of Medical Science*, **146**, 307–308.

Wickstrom, G., 1978. Symptoms and signs of degenerative back disease in concrete reinforcement workers. *Scandinavian Journal of Work, Environment, and Health*, **4** (Suppl. 1) 54–58.

Wyndham, C.H., Strydom, N.B., Morrison, J.F., Williams, C.G., Bredell, G., Peter, J., Cokke, H.M. and Joffe, A., 1963. The influence of gross weight on oxygen consumption and on physical working capacity of manual laborers. *Ergonomics*, **6**, 275–286.

Yates, J.W., Kamon, E., Rodgers, S.H. and Champney, P.C., 1980. Static lifting strength and maximal isometric voluntary contractions of back, arm and shoulder muscles. *Ergonomics*, **23**, 37–47.

Chapter 3

Design approaches to solving manual materials handling problems

Introduction

Workers engaged in physical work are affected by forces from the immediate physical environment as well as the biomechanical forces generated within the body. The net forces and/or effort imposed upon the worker produce a strain on the worker's musculoskeletal and physiological systems. The musculoskeletal and physiological systems and their response to the demands of a task are described by the biological laws of the body. The goal of ergonomists is to reduce the stress imposed upon the body sufficiently to minimize musculoskeletal and/or physiological strain (Ayoub and Mital, 1989).

In the context of MMH, a variety of approaches have been used to quantify the relationships between imposed stresses and the resulting strain and, thereby, control the pervasive overexertion injury and back problem. Specifically, the following approaches have been developed:

(1) the epidemiological approach
(2) the biomechanical approach
(3) the physiological approach
(4) the psychophysical approach

All these approaches to designing MMH tasks are diverse and seemingly unrelated, and yet have been practised for a considerable length of time. The epidemiological approach investigates the circumstances and conditions that exist during an incident and attempts to develop a set of general conditions that may be associated with the hazards of MMH activities. The biomechanical approach aims at determining the tissue tolerance characteristics of the spinal column, spinal shrinkage, spinal compression or the pressure generated in the abdominal cavity, the intra-abdominal pressure (IAP). The physiological approach relates the metabolic and circulatory costs of performing an MMH task to the workers' physiological limit. The psychophysical approach relies on the individual to quantify his/her stress tolerance level and establishes the acceptable lifting weights or forces. Following is a brief description of each of the design approaches and the specific design criterion used for each approach and its limitations. Finally, the various design approaches are compared and an attempt is made to develop a consensus to provide a margin of safety for the human back during MMH activities.

The epidemiological approach

'Epidemiology is the study of disease occurrence in human populations' (Friedman, 1974). The emphasis is on groups of people rather than the individual. In the context of MMH activities, epidemiology attempts to answer questions such as 'Why are there so many back injuries in the material handling occupation?', 'How can these injuries be prevented?', 'What are the common factors present in all or most of these injuries?', 'What factors are more likely to predict future overexertion injuries?', and 'How does the probability of injury vary with changes in one or more of the factors considered to be risk factors?'. In general, epidemiology is concerned with discerning the injury patterns present, if any, and using these patterns to predict the occurrence of injury.

The basic measurements in epidemiology are: counts (number of people in group suffering from back injuries, a particular back disorder, low-back pain, etc.), prevalence rate (number of people in a group inflicted with some back disorder/total number of people in the group), and incidence rate ((number of people developing a disorder/total number at risk)/unit time). These basic measurements can be used to compare occurrence of disorder among different groups. The basic measurements and the comparison of these factors for groups of interest can identify factors that may be considered as risk factors in MMH activities.

The injury and back disorder risk factors in MMH activities are broadly grouped in two categories: (1) personal risk factors and (2) workplace risk factors (Andersson, 1981). Many of the factors that are considered risk factors have been reviewed in Chapter 2. The detailed review on these and other factors, such as asymmetrical handling, is provided in Ayoub and Mital (1989) and the reader is referred to this reference for a detailed literature review.

A number of personal and workplace factors have been identified as possible risk factors. The most significant of these factors are listed in Table 3.1. However, epidemiological evidence relating these factors to back injuries and back-related disorders is far from convincing or definitive. Since most of these factors affect physiological and musculoskeletal systems and, thereby, influence MMH capability of people and also frequently appear in many MMH capability prediction models, it is logical to expect these factors to be correlated with injuries and back disorders. Unfortunately, the epidemiological evidence supporting a link between most of these factors and injury or back disorders is either unavailable or the available data support only a qualitative link. Very few risk factors have been definitively linked to injuries and back disorders.

Table 3.1 MMH-related injury and back disorder risk factors

Personal risk factors	Workplace risk factors
Physique/anthropometry/strength	Posture/handling techniques
Physical fitness/spinal mobility	Load characteristics (weight, size, shape, couplings, etc.)
Age/gender	Handles/grip
Psychophysical factors/motivation	Repetitive handling
Training and selection	Confined environments/spatial restraints
Static/dynamic endurance	Safety aspects
Health history/spinal abnormalities	Protective equipment
Experience	Work organization/workplace geometry
	Environment (heat, humidity, noise, glare, etc.)
	Asymmetrical handling (twisting and turning)
	Reaching/stretching
	Task duration

Several epidemiological studies have concluded that weight of the load and frequency of lifting it are among the most critical factors that determine the risk of injury. Chaffin and Park (1973) reported that the lifting strength ratio of the individual was highly correlated with the incidence rate of low-back pain (Figure 3.1). Chaffin *et al.* (1976) reported that the heavier the loads lifted, the greater was the severity

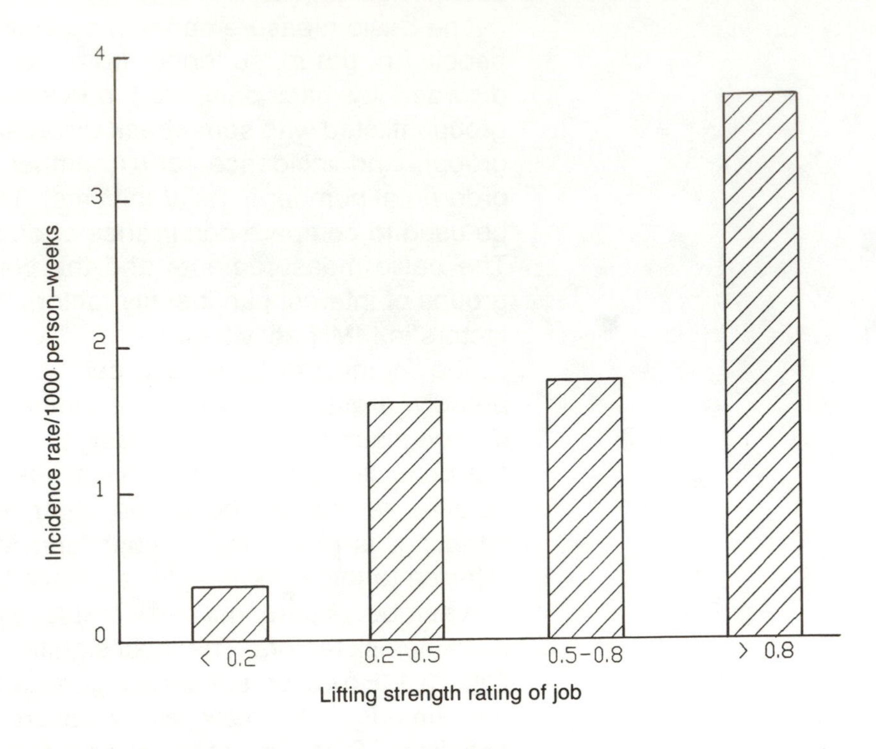

Figure 3.1 Incidence rates of back complaints per 1000 person-weeks and lifting strength rating
(data from Chaffin and Park, 1973; from Jensen, 1988).

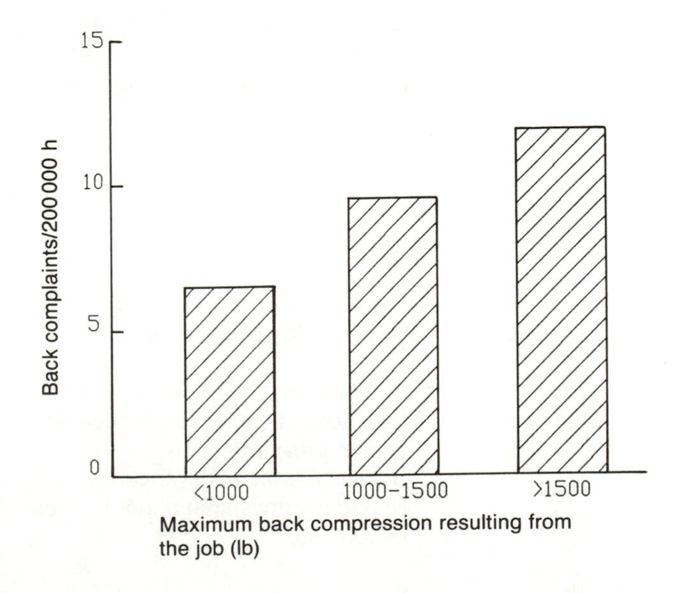

Figure 3.2 Back complaints (per 200 000 h) versus compressive force on the L5/S1 disc
(data from Herrin et al., *1986; from Jensen, 1988).*

of injuries as indicated by lost or medically restricted workdays. They further concluded that the frequency and severity rates of musculoskeletal problems in areas other than the back and the severity of contact injuries increased with the frequency of lifting maximal loads on the job. The authors also found a strong correlation between the compressive force on the L5/S1 disc and the incidence rate of low-back pain. They observed that jobs that exceeded the compressive force of 6236 N had an eight times greater low-back pain incidence rate than jobs that had a compressive force less than 2673 N. This finding was further confirmed by Herrin *et al*. (1986); they reported that the incidence of back disorders increased as the compressive force on the L5/S1 disc increased (Figure 3.2). As stated above, the compressive force determination was based on static strength models.

Chaffin *et al*. (1978) found that back injury severity and incidence rate increased almost three-fold as the job lifting requirements approached lifting strength of workers (Figure 3.3). Ayoub *et al*. (1983), using the Job Stress Index (JSI; Ayoub and Mital, 1989; a working duration, lifting frequency, load and lifting capacity based index) concept, reported a critical JSI value of 1.5; the occurrence of injuries increases substantially when the JSI is above 1.5. The study mainly included male workers. Figure 3.4 describes the relationship between JSI and cumulative injury rate. A substantial increase in cumulative injury rate is evident beyond the JSI value of 1.5. The maximum weight of the load associated with the JSI value of 1.5 was 60 lb (27.24 kg). For this load, the compressive force on the spine is approximately 3930 N. **It is important to note that JSI considers the dynamics and fatigue of the job, and the calculation of spinal compression is based on a three-dimensional dynamic biomechanical model (Kromodihardjo and Mital, 1986, 1987).**

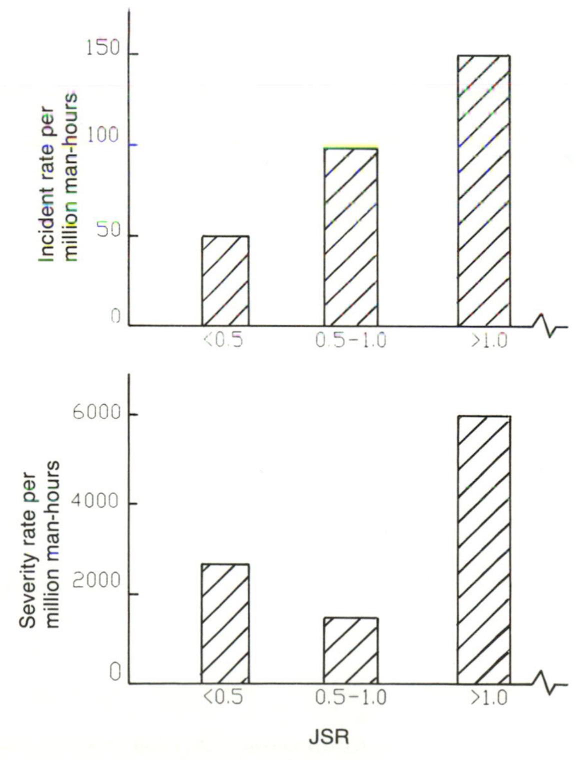

Figure 3.3 Back injuries versus job strength rating (Chaffin et al.*, 1978).*

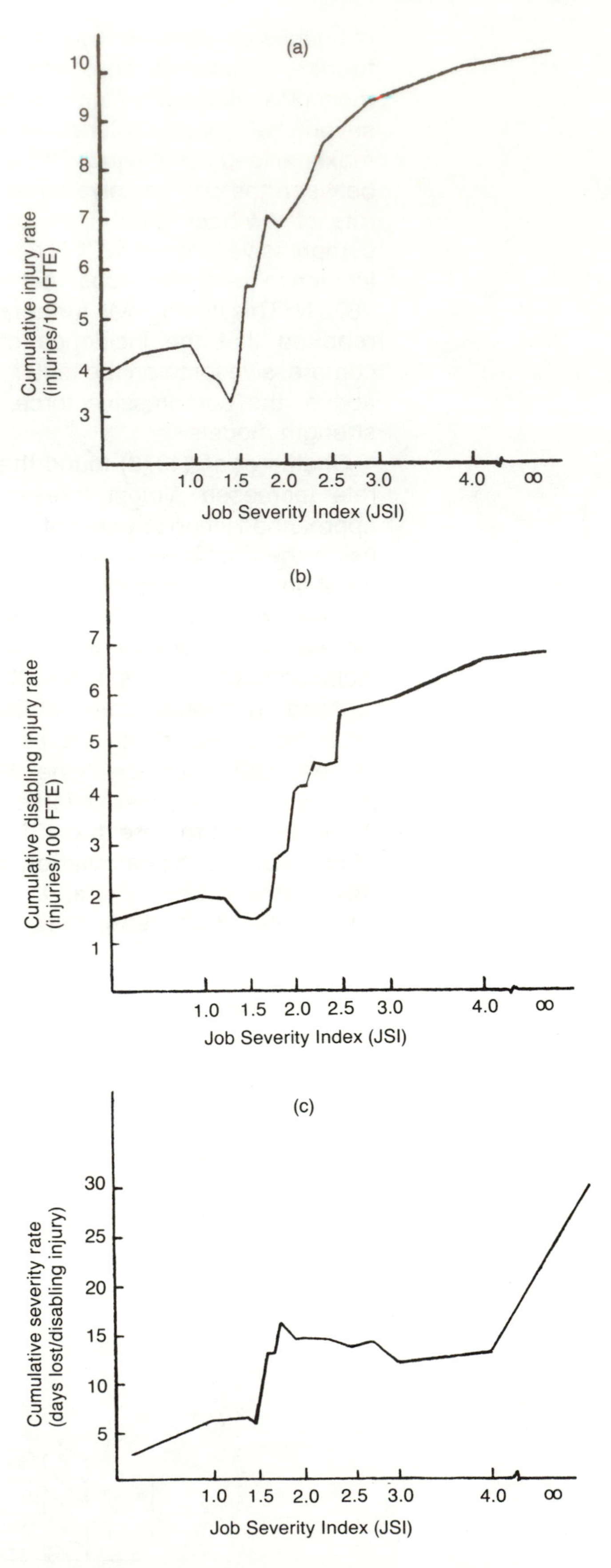

Figure 3.4 Relationship between JSI and (a) cumulative injury rate, (b) cumulative disabling injury rate, and (c) cumulative severity rate (Ayoub et al., *1983).*

FTE, full-time equivalent.

The spinal compression of 3930 N is approximately 69% of the mean ultimate compression strength of male lumbar spine elements (5700 N; Jager and Luttmann, 1991). If the same proportion is applied to females, the spinal compression limit for females would be 2689 N. This would correspond to a weight of 20 kg.

Frequency of lifting has also been positively linked to low-back pain (Frymoyer *et al.*, 1980; Arad and Ryan, 1986). Frymoyer *et al.* (1983) maintain that lifting a load in excess of 20 kg, repetitively, is the most important prognosis variable for low-back pain. Kelsey *et al.* (1984) concluded that the risk of acute prolapsed lumbar intervertebral disc was three times higher in jobs that required lifting more than 25 lb in excess of 25 times per day as compared to jobs that required lifting less than 25 lb. Damkot *et al.* (1984) also made a qualitative finding that heavier loads were lifted by individuals complaining of severe low-back pain as opposed to individuals who complained of moderate or no pain.

Many other studies have also reported a qualitative relationship between other workplace-related risk factors and MMH injuries. For instance, Snook *et al.* (1978) reported that almost 18% of all back injuries were due to twisting and turning. They also observed that two-thirds of the lifting tasks involved bending and accounted for 78% of all lifting injuries. An association between bending and low-back pain has also been reported by Takala and Kukkonen (1987). Stretching, reaching, pushing and pulling have also been implicated in severe low-back pain incidences (Damkot *et al.*, 1984).

Several studies have reported that individuals with a previous history of back or sciatic pain have greater prevalence of back pain and prolonged disability than those who have had no prior such history (Lawrence, 1955; Chaffin and Park, 1973; Bergquist-Ullman and Larson, 1977; Troup *et al.*, 1981; Nachemson, 1982; Rowe, 1983). It is now generally agreed that a person who has had back or sciatic pain in the past is more likely to have it in the future than a person who has never had it. **Therefore, good physical histories of individuals being considered for MMH activities are essential.**

As this brief review of epidemiological studies shows, much of the information linking the risk factors, personal or workplace-related, and injury and back disorders is qualitative. Even where quantitative data are available, the risk factor that might predict future injury does not account for all stresses and hazards. For instance, the relationship between back incidence and lifting/job strength ratings; these strengths not only do not account for fatigue, they also do not account for other hazards such as repetition and duration.

Epidemiological studies also suffer from many other difficulties. Most studies have small samples, lack a control, are conducted for short periods, and are retrospective. Dealing with subjective information, such as complaints of low-back pain, which frequently may be unreliable is another major shortcoming of such studies (Glover, 1970).

The design criteria

Given the lack of reliable quantitative relationships between various risk factors and injury and back disorders, it is very difficult to come up with specific quantitative criteria that may be used in designing MMH jobs. Some guidelines are, however, possible. **MMH jobs that yield a JSI (Ayoub and Mital, 1989) in excess of 1.5 are almost certain to**

cause a substantial increase in severe injuries. Jobs that cause a compressive force greater than 3930 N on the L5/S1 disc of males are also likely to trigger back complaints; jobs that lead to compressive forces in excess of 6236 N (determined from static strength models) most certainly will lead to back incidences. For females, jobs causing a spinal compression in excess of 2689 N are likely to trigger back problems.

The biomechanical approach

The biomechanical approach to estimate the mechanical stresses on the body (primarily forces acting on the lower back) relies on two measures: (1) the compression and shear forces generated at the L5/S1 disc of the spinal column and (2) pressures generated in the abdominal cavity (IAP). The first measure of stress requires that the human body be treated as a system of links and connecting joints. A variety of models have been developed over the years to evaluate industrial MMH tasks, manual lifting tasks in particular (Ayoub and Mital, 1989). In the biomechanical model, each of the links is the same length and possesses the same mass and moment of inertia as their corresponding human segments. The mass is considered to be concentrated at a single point on the link, the centre of mass.

The various biomechanical models available in the published literature differ primarily in two ways: (1) the number of links and (2) the analysis technique (two-dimensional versus three-dimensional). In general, most two-dimensional biomechanical models have five to seven links; the number of links in a three-dimensional model is considerably more (Kromodihardjo and Mital, 1986, 1987).

In order to estimate the mechanical stress imposed on the body while at rest or in motion, the various mechanical properties of the body segments to perform the appropriate analysis are used. A number of simplifications and assumptions concerning the human body are necessary. The nature and extent of these assumptions and simplifications determine the sophistication of a biomechanical model. Using the various internal and external forces (gravitational forces, ground reaction or external forces, and muscle forces) the human body is subjected to during an MMH activity, the biomechanical model computes the forces (compressive and shear) on the L5/S1 disc of the spinal column. The L5/S1 disc has been identified as the weakest link in the body segment chain (Ayoub and Mital, 1989). The compression force on the L5/S1 disc, as a result of performing an MMH activity, is compared with the compressive strength of the spinal column (intervertebral discs and vertebrae) (the compression strength of the spinal column factors that affect this strength and limitations of this approach are discussed further under the next section). The factors that affect spinal compression and shear forces have been discussed by Ayoub and Mital (1989) and the reader is referred to this reference for details.

The second measure of stress requires measurement of the IAP. When loads are lifted, muscles in the lower back are stressed and pressures are generated in the abdominal cavity (Davis *et al*., 1965). It has been observed that occupations that lead to an IAP above 100 mm Hg have a significantly high incidence of low-back pain (Davis and Stubbs, 1978; Mairiaux *et al*., 1984; Davis, 1985). There is, however, a group of researchers who believe that the IAP generated during MMH activities helps to relieve some part of the load applied on the spine by producing an extension moment. Morris *et al.* (1961), for instance, maintain that the calculated compressive force of about 30% on the lumbosacral level and about 50% on the lower thoracic portion could be sustained by this IAP during lifting a load. Many other studies have also indicated similar findings (Eie, 1966; Schultz *et al.*, 1982; Thomson, 1988).

This role of IAP, reducing the compressive force on the spinal column, however, is currently surrounded in controversy. Several researchers maintain that the abdominal cavity diaphragm area is inadequate to generate sufficient IAP to alleviate spinal compression (Leskinen *et al.*, 1983a, b). Others maintain that the net effect of extensor activity and flexor moment, due to increased abdominal muscle activity due to IAP, results in a net increase in spinal compression (McGill and Norman, 1985; Nachemson *et al.*, 1986).

Given the above controversy regarding the impact of IAP on spinal compression, it is difficult to comment on its role in either alleviating or adding to spinal compression. There are, as indicated earlier, several studies that have correlated the incidence of low-back pain with alleviated IAP. From that standpoint alone, IAP merits serious consideration as a criterion in designing MMH activities. The effect of various task factors on IAP is discussed in the next section.

The design criteria

The design criterion which relies on the measurement of compression and shear forces at the L5/S1 disc of the lumbar spine requires that these spinal forces be compared with the load tolerance capability of the lumbar spinal column. However, the compressive strength of the lumbar spinal column appears to be the only strength that has been widely used as the design criterion. A discussion of the compressive strength of the lumbar spinal column is presented below.

Several researchers have looked into the material properties of the spinal structure (Bartlink, 1957; Perey, 1957; Evans and Lissner, 1959; Sonoda, 1962; Hutton and Adams, 1982; Brinckmann *et al.*, 1987, 1988). From these studies it appears that the ultimate compressive strength of the lumbar spinal column varies between 3000 N and 12 000 N. Evidently, there is a great deal of variability in the compressive strength values. Furthermore, the compressive strength of the lumbar spinal column is affected by a variety of factors. Table 3.2 lists these factors.

Jager and Luttman (1991) integrated the results of several studies aimed at determining the compressive strength of the lumbar spinal column. Their findings revealed that on average, male lumbar spine elements fail at a compression of 5700 N (standard deviation = 2600 N). For female spinal column elements, the failure due to compression occurs at 3900 N (standard deviation = 1500 N). Furthermore, the factors that definitely influence the compressive strength of the spinal column are: age (correlation coefficient = −0.48), gender (correlation coefficient = 0.37), specimen cross-section (correlation coefficient = 0.28), lumbar level (correlation coefficient = −0.27) and structure (correlation coefficient = −0.20). According to Jager and Luttmann (1991), the ultimate compressive strength of the spinal column could be determined from the following relationship:

$$\text{compressive strength (kN)} = (7.65 + 1.18G) - (0.502 + 0.382G)A + (0.035 + 0.127G)C - 0.167L - 0.89S$$

Table 3.2 Factors affecting compressive strength of the lumbar spinal column

Factor	Study
Age	Adams and Hutton (1982), Biggemann *et al.* (1988), Brinckmann *et al.* (1988, 1989), Hansson *et al.* (1980, 1987), Hutton and Adams (1982), Hutton *et al.* (1979)
Gender	Adams and Hutton (1982), Biggemann *et al.* (1988), Brinckmann *et al.* (1988, 1989), Hansson *et al.* (1980), Hutton and Adams (1982), Hutton *et al.* (1979)
Body weight	Adams and Hutton (1982), Hansson *et al.* (1987), Hutton and Adams (1982)
Spinal level	Sonoda (1962), Yamada (1970)
Spinal components	Adams and Hutton (1982), Biggemann *et al.* (1988), Brinckmann *et al.* (1988, 1989), Hansson *et al.* (1980), Hutton and Adams (1982), Sonoda (1962)
Loading	Brown *et al.* (1957), Evans and Lissner (1959), Lin *et al.* (1978)
Posture	Adams and Hutton (1982, 1985), Hutton and Adams (1982)
Physical activity	Porter *et al.* (1989)
Type of specimen	Evans and Lissner (1959)
Cortical shell/trabecular	McBroom *et al.* (1985)

where G is gender (0 for female; 1 for male), A is in decade (e.g. 30 years = 3, 60 years = 6), L is lumbar level (0 for L5–S1; incremental value for each lumbar disc or vertebra), C is cross-section (cm^2), and S is structure (0 for disc; 1 for vertebra). Thus, for example, the predicted compressive strength for the L2 vertebra with an area of 18 cm^2, taken from a 30-year-old man will be 7 kN.

Given the wide variation in the ultimate compression of the lumbar spinal column and the fact that it is affected by a variety of factors and, therefore, would vary from person-to-person, it is difficult to come up with a specific value as the design criterion. This difficulty has been circumvented to a certain extent by simply considering critical limits of compressive strength based on JSI (Ayoub *et al.*, 1983; see Figure 3.4) and compressive force–back incidence rate data (Chaffin and Park, 1973; see Figure 3.5). According to these data, **any industrial job creating 650 kg (6374 N) compressive force on the spine will put workers in the over 35 years age group at a greater risk of low-back injuries. Compression forces of approximately 400 kg (3930 N) could be tolerated by most males; for females, compression forces of approximately 274 kg (2689 N) could be tolerated (see discussion under the epidemiological approach).**

It should be kept in mind that there is evidence (Brinckmann *et al.*, 1988; Jager and Luttmann, 1991) to suggest that even at the compression limits suggested above, a significant proportion of the population (20%) may be at risk, particularly those above 60 years of age. Furthermore, it should also be realized that compression strength should not be the sole design criterion. This is particularly true when the MMH task requires asymmetrical

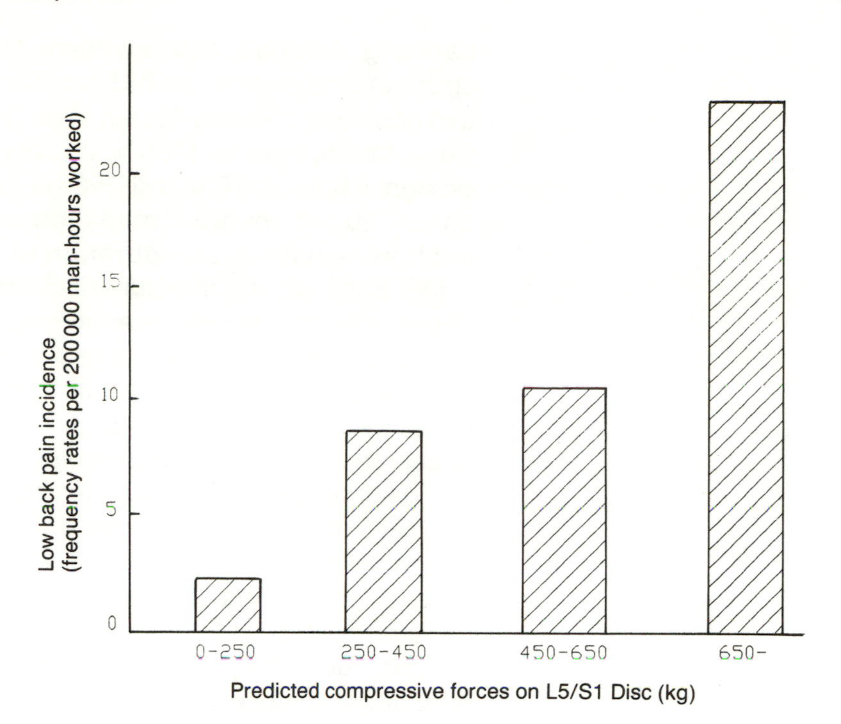

Figure 3.5 Relation between low-back pain and compressive force (Chaffin and Park, 1973).

or repetitive handling. The studies investigating asymmetrical handling are relatively few, but do indicate that asymmetrical lifting is much more stressful (Kumar, 1980, 1984; Mital and Kromodihardjo, 1986). The work of Mital and Kromodihardjo (1986) also indicates that spinal column shear strength may be relatively more critical in such cases than its compressive strength (Figure 3.6).

However, in spite of the realization that asymmetrical handling activities appear to be relatively more hazardous than symmetrical

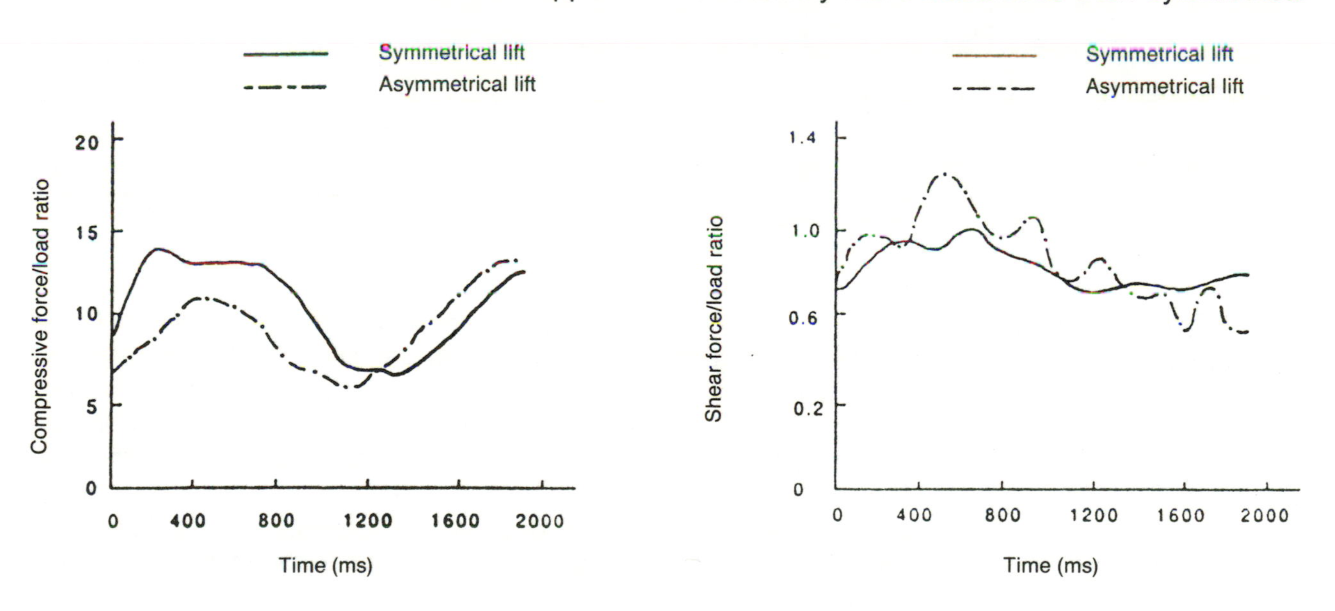

Figure 3.6 Normalized compressive and shear forces for symmetric and asymmetric lifting (Mital and Kromodihardjo, 1986).

handling activities, establishment of a lumbar spinal column shear strength design criterion has been overlooked. The only reference to this has been made by Farfan *et al.* (1970) who recommended **a shear force failure limit of 1735 N. It is also worth noting that both these design limits (3930 N and 2689 N for compression and 1735 N for shear) do not consider many important work-related risk factors, such as repetition and duration of handling.**

IAP is the second measure of MMH stress. As indicated earlier, IAP above 100 mm Hg has been strongly correlated to severe low-back pain incidences in occupations requiring MMH. **Using 90 mm Hg as the upper limit of IAP that could be sustained without the risk of back disorders,** Davis and Stubbs (1980) developed contours of acceptable force limits for a variety of MMH activities. Although the IAP design criterion (90 mm Hg) does consider a variety of MMH activities, it makes a sweeping adjustment for repetitive handling by simply suggesting a reduction of 30% in force recommendations for infrequent handling. There are many studies that have investigated the effect of repetition on MMH capacity and concluded that a simple adjustment factor does not suffice; the increase in frequency of handling from once a minute to 12 times per minute could reduce MMH capacity by nearly 30% (Ayoub and Mital, 1989). Table 3.3 shows the effect of lifting frequency on lifting capacity based on the study by Mital (1984a, b). It has also been reported by Legg (1981) that IAP is not sensitive to fatigue or training of abdominal muscles. However, it is sensitive to the weight of the load and activity (Andersson *et al.*, 1979).

Table 3.3 Effect of lifting frequency on acceptable lifting capacity

Shift (h)	Lifting frequency (lifts/min)	Decline in acceptable lifting capacity (%) from 1 lift/min
8	1	0
	4	7.5
	8	18
	12	25.5
12	1	0
	4	8.5
	8	19.5
	12	27.5

(Mital, 1984a,b)

The physiological approach

Unlike the biomechanical design approach which primarily applies to infrequent lifting, the physiological approach is applicable to repetitive lifting where the load is within the physical strength of the worker. During repetitive handling tasks, a person's endurance is primarily limited by the capacity of the oxygen transport system. As muscles contract and relax, their increased metabolic energy demand requires an increase in the delivery of oxygen and nutrients to the tissue. If this demand for increased oxygen and nutrients cannot be met, the activity cannot continue for long.

When a person is engaged in physical work, such as MMH activities, several physiological responses are affected. These include metabolic energy cost, heart rate, blood pressure, blood lactate and ventilation volume. Of all these responses, metabolic energy expenditure has been the widely accepted physiological response to repetitive handling as it is directly proportional to the workload at steady-state conditions (Durnin and Passmore, 1967; Aquilano, 1968; Hamilton and Chase, 1969; Ayoub *et al*., 1981; Mital, 1984a, b; Astrand and Rodahl, 1986). For this reason, our discussion will exclusively focus on metabolic energy expenditure rate as the physiological approach design criterion.

Several work- and workplace-related factors affect metabolic energy expenditure rate. Table 3.4 summarizes these factors and their net effect on oxygen consumption. For a detailed discussion on the effect of these and personal factors on oxygen consumption the reader is referred to *Manual Materials Handling* by Ayoub and Mital (1989).

Table 3.4 Net effect of work and workplace factors on metabolic energy

Factor	MMH activity	Net effect
Frequency of handling (↑)	All	Increase
Task duration (↑)	All	Increase[a]/Decrease[b]
Object size (↑)	All	Increase
Couplings (good)	All	Decrease
Object shape (various)	All	Unknown
Object weight/force (↑)	All	Increase
Load stability/distribution	Lifting,carrying	Unknown
Vertical height (↑)	Lifting, lowering	Increase
Distance travelled (↑)	Pushing, pulling, carrying	Increase
Speed/grade (↑)	Pushing, pulling, carrying	Increase
Asymmetrical handling	Lifting	None

↑ increase; [a] if the weight/force does not change; [b] if the weight/force decreases (e.g. when using the psychophysical methodology).

As Table 3.4 shows, the metabolic energy expenditure is influenced by a very wide variety of risk factors. The effects of other factors on metabolic energy, such as posture, though too involved to be included in this table, are nevertheless also significant. The primary reason for the sensitivity of metabolic energy expenditure rate to work-related risk factors is the fact that energy cost is dependent upon the amount of muscle groups active during task performance. Since all work factors, including those listed in Table 3.3, affect the amount of muscle groups, the metabolic energy expenditure varies with them.

In addition to work- and workplace-related risk factors, there are personal and environmental factors that also influence oxygen consumption. A detailed discussion on the effect of these factors on metabolic energy expenditure is provided by Ayoub and Mital (1989).

The design criterion

The physiological approach expresses job stresses as metabolic energy expenditure requirements. The maximum metabolic energy expenditure rate that can be sustained without overexertion and excessive fatigue determines the design criterion limit. In order to determine the design criterion limit, two basic questions must be answered:

1. What is the upper limit of oxygen consumption, expressed as a percentage of aerobic capacity, that can be sustained without undue fatigue while performing MMH activities?
2. What kind of aerobic capacity should be used to express this percentage?

Several investigations have attempted to answer the first question. Muller (1953) and Bink (1962) proposed an upper limit of 5.0–5.2 kcal/min (5 kcal/min = 1 l of oxygen/min) for daily work. Michael *et al.* (1961) concluded that 35% of the maximum aerobic capacity (bicycle ergometry and treadmill walking) was the limit of physical work that could be performed without undue fatigue. Astrand (1967) reported that 50% of bicycle aerobic capacity could not be sustained by all individuals, thus putting an absolute upper limit on acceptable levels of oxygen consumption in daily work. Petrofsky and Lind (1978) determined that 25% of bicycle aerobic capacity could be sustained without fatigue. They also determined that aerobic capacity is task specific. They further determined that lifting aerobic capacity was lower than bicycle aerobic capacity.

While the above recommendations differ considerably, it has been generally agreed that physical jobs that require a metabolic energy of more than 5 kcal/min (approximately 33–35% of the treadmill aerobic capacity; Michael *et al.*, 1961) will lead to overexertion and undue fatigue. This physiological fatigue guideline was tested by Mital (1984a). He reported that industrial male workers, when lifting psychophysically acceptable loads across a wide frequency range for 8 h, selected weights that on average resulted in a metabolic energy expenditure rate of 28% of their bicycle aerobic capacity; industrial females selected weights corresponding to 29% of their bicycle aerobic capacity. When the working duration increased from 8 h to 12 h, the aerobic capacity percentages went down to 23% and 24% for males and females, respectively. Legg and Pateman (1985) and Legg and Myles (1985) also reported that their subjects lifted loads that required 21% of subjects' uphill treadmill aerobic capacity (uphill treadmill aerobic capacity is higher than bicycle aerobic capacity which, in turn, is generally higher than lifting aerobic capacity; Sharp *et al.*, 1988). Since the heart rates in the studies by both Mital (1984a) and Legg and Myles (1985) were very close, it can be taken that the two recommendations are very similar (people would require a lower percentage of uphill treadmill aerobic capacity than bicycle aerobic capacity; in absolute terms (litres of oxygen per minute) the two will lead to similar values for MMH tasks).

Legg and Pateman (1985) and Legg and Myles (1985) also argue that since MMH activities have a combination of static and dynamic components, **21–23% of uphill treadmill aerobic capacity should be considered as the upper limit of metabolic energy expenditure that could be sustained for any load/repetition combination for 8-h workdays. In terms of bicycle aerobic capacity, the recommendation would be approximately 28–29%. This limit will have to be revised downwards, to 23–24% of bicycle aerobic capacity, for 12-h workdays.**

The second question that was posed in the context of using the physiological approach for designing MMH jobs was 'What kind of aerobic capacity should be used?'. As indicated by Petrofsky and Lind (1978), aerobic capacity is task dependent. Khalil *et al.* (1985), Fernandez (1985), Kim (1990) and Mital *et al.* (1987a) also determined that lifting aerobic capacity is a function of load and frequency. Since, in a work situation there could be numerous load–frequency combinations, if we just consider the effect of these two work factors, designing MMH jobs on the basis of these different aerobic capacities would be very tedious. To circumvent this problem, **uphill treadmill or bicycle aerobic capacity should be used. The criterion limit would**

be 21–23% of uphill treadmill aerobic capacity or 28–29% bicycle aerobic capacity for an 8-h workday.

The psychophysical approach

The psychophysical approach to MMH job design requires individuals to adjust either the handling frequency, the weight of the load or the force exerted on the object being handled according to their perception of physical strain. All variables other than the one being adjusted are controlled. The individuals adjust their workload (kilogram-metre/minute) to the maximum amount they can sustain without undue strain or discomfort, and without becoming unusually tired, weakened or overheated, or out of breath. The final workload is the maximum acceptable frequency of handling or the maximum acceptable weight/force of handling.

The psychophysical methodology has been successfully used by a number of researchers and has a major advantage over other design approaches in that it can be used for both frequent as well as infrequent MMH activities. The major disadvantage of this approach is that the need for special control restricts its use primarily to laboratory investigations.

The methodology requires that individuals be started randomly with either a very light or heavy load/force (or low or high frequency if frequency is being adjusted) and they are allowed to adjust it until they arrive at the workload that can be sustained for the projected working duration without discomfort or undue fatigue. The final weight/force or frequency is recorded and is used in determining the criterion limit for the individual or a specific population. The process is repeated and, if the two trials are within 15% of each other, the results are averaged (Snook, 1978).

A number of personal, work and environmental factors affect the psychophysical design criterion. The details of how these factors affect acceptable weight/force are given by Ayoub and Mital (1989). Table 3.5 summarizes the net effect of some of the important work factors.

Table 3.5 Net effect of work-related factors on acceptable weight/force

Factor	MMH activity	Net effect
Frequency (↑)	All	Decrease
Task duration (↑)	All	Decrease
Object size (↑)	All	Decrease
Object shape (various)		
Collapsible (e.g. bags)	Lifting,carrying	Increase
Non-collapsible (e.g. metallic) (Volume increases)		Increase
Non-collapsible (Volume does not change)		Unknown
Couplings (Good)	All	Increase
Load stability/distribution	Lifting,carrying	Decrease
Vertical lift height (↑)	Lifting, lowering	Decrease
Height of force (↑)	Pulling, pushing	Increase
Application/starting point	Lifting, lowering, carrying	Decrease
Distance travelled (↑)	Pushing, pulling, carrying	Decrease
Speed/grade (↑)	Pushing, pulling, carrying	Decrease
Asymmetrical handling	Lifting, lowering	Decrease

↑ increase

The design criteria

The psychophysical design approach has two design criteria: (1) maximum acceptable frequency of handling, and (2) maximum acceptable weight/force of handling. In most psychophysical studies, the second criterion has been used. Nevertheless, it is pertinent to discuss the first criterion as well.

Relatively few studies have determined the maximum acceptable frequency of handling. All have dealt with either two-handed lifting (Snook and Ciriello, 1974; Snook and Irvine, 1968; Mital *et al.*, 1987b) or one-handed lifting (Garg and Saxena, 1982; Mital and Asfour, 1983). **The maximum acceptable frequency of lift for one-handed lifting tasks has been determined to be 50% of the frequency that can be sustained over 4 min.** Snook and Ciriello (1974) reported that for two-handed lifting tasks the maximum acceptable frequency of lift is 4 lifts/min. This was not the choice of workers when they were allowed to determine the pace of stacking loaded cartons on a pallet (Mital *et al.*, 1987b). On the average, males chose a frequency of 12.9 lifts/min while females chose 9.7 lifts/min. The metabolic cost of stacking the loads on the pallet were, however, higher than the physiological design criterion ceiling. When the frequency was reduced to a fixed level of 6 lifts/min, the metabolic cost for males dropped to within acceptable levels. Females, even at the frequency of 6 lifts/min, exceeded the metabolic cost ceiling. It is worth noting that the maximum acceptable frequency of lift is heavily dependent on the weight of the load being lifted; the heavier the load, the lower the acceptable frequency of lift is expected to be. **From these studies, one could draw a general conclusion that industrial workers, when given the choice, select a lifting frequency between 4 and 6 lifts/min.**

Several studies have been undertaken to determine the maximum acceptable weight/force of handling for various MMH activities. The ceilings for this design criterion have been reviewed in detail by Ayoub and Mital (1989). The two main studies that could not be included are by Ciriello *et al.* (1990) and Mital (1992) and are considered in developing the overall recommendations that are provided in Part II of this guide.

Comparison of design approaches

Historically, researchers have believed that the various MMH design approaches are different and lead to different recommendations for load, or force, that can be sustained without undue risk of injury or fatigue. This belief is based primarily on the following two facts: (1) different design approaches employ different design criteria and (2) the conditions under which some of the design approaches can be used differ; the biomechanical approach, for instance, is primarily limited to infrequent handling. Garg and Ayoub (1980) provided further support to this notion when they compared load recommendations available in the published literature for manual lifting activities. They concluded that: (1) the recommendations provided by different approaches are not in agreement, (2) the maximum acceptable weights of the load based on psychophysical studies are lower than those based on biomechanical criterion (isometric strength), and (3) the psychophysical design criterion (maximum acceptable weight of lift) leads to greater work loads at higher frequencies when compared to the physiological design criterion of 5 kcal/min (this trend is reversed at lower frequencies; Figure 3.7). It is worth noting that this comparison was very limited as it did not include all the design criteria

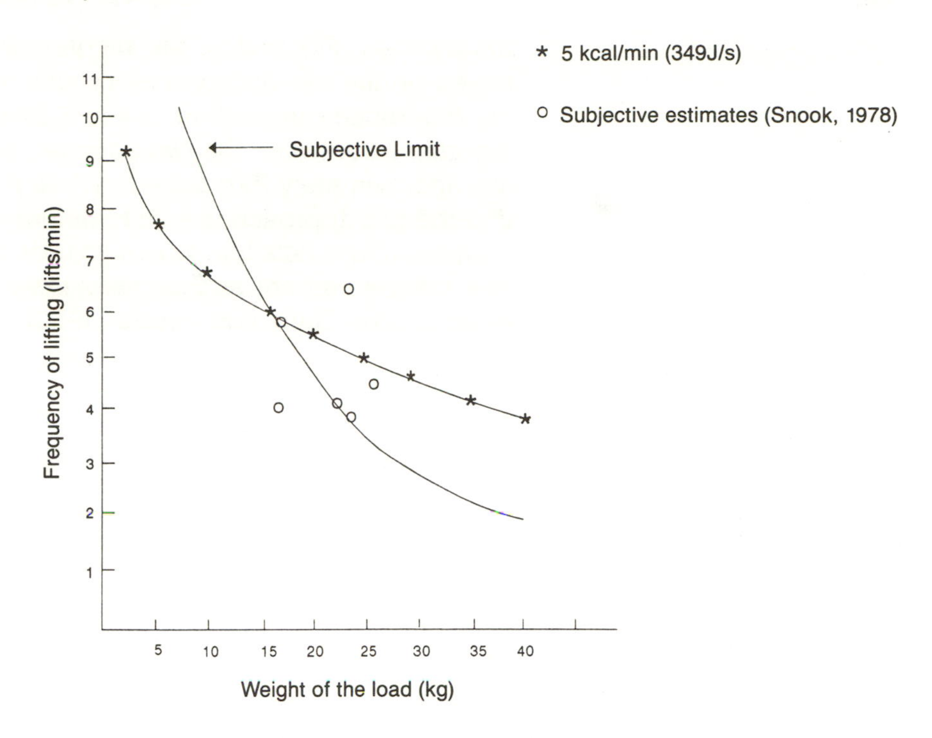

Figure 3.7 Comparison of psychophysical and physiological weight recommendations for lifting from floor to 0.51 m height (Garg and Ayoub, 1980).

discussed earlier. The ceiling of the design criterion (metabolic energy expenditure rate) was higher than what has been proposed in the last few years. Furthermore, the comparison relied on data gathered from different studies; the experimental conditions and control quite frequently vary considerably from study-to-study and no effort was made to account for it.

Subsequent to the report by Garg and Ayoub (1980), Ciriello and Snook (1983) and Mital (1983) reported that psychophysical methodology overestimates manual lifting capabilities of workers by as much as 30%. It was further reported by Mital (1984a, b) that the metabolic energy expenditure criterion of one-third aerobic capacity leads to metabolic overloading and must be downgraded if overexertion is to be avoided. A limit of 29% of aerobic capacity (0.8 l/min) for males and 28% of aerobic capacity (0.6 l/min) for females was proposed. Mital (1985) compared the safe weight recommendations based on the psychophysical approach, proposed criterion ceilings for the physiological approach, and the earlier criterion ceiling of one-third aerobic capacity across low and high lifting frequencies. Experienced industrial workers were employed and physiological and psychophysical data were collected in a single experiment. Figures 3.8 and 3.9 show the relationship between weight, oxygen uptake and frequency for males and females, respectively. The points of intersection of frequency and criteria lines (Figures 3.8 and 3.9) were projected vertically downwards (on the weight axis) to determine physiologically acceptable weights. These weights and psychophysically acceptable weights were plotted with frequency (Figures 3.10 and 3.11). Both Figures 3.10 and 3.11 show that weight recommendations based on the physiological approach are consistently higher than the psychophysical weight estimates. The differences are greater at lower frequencies than at higher frequencies. For males, the difference in weight recommendations based on the two approaches ranges from 0.3 to 4.3 kg; for females the difference ranges from 0.4 to 5.5 kg. At 4 lifts/min, or higher, the differences between the two approaches for both males and females are approximately 2 kg or less. While Figures 3.10 and 3.11 do show that the two approaches lead to different weight recommendations, it is interesting to note that the magnitude of differences is not very large. The differences are substantially lower and different than for those projected by Garg and Ayoub (1980). For instance, based on data

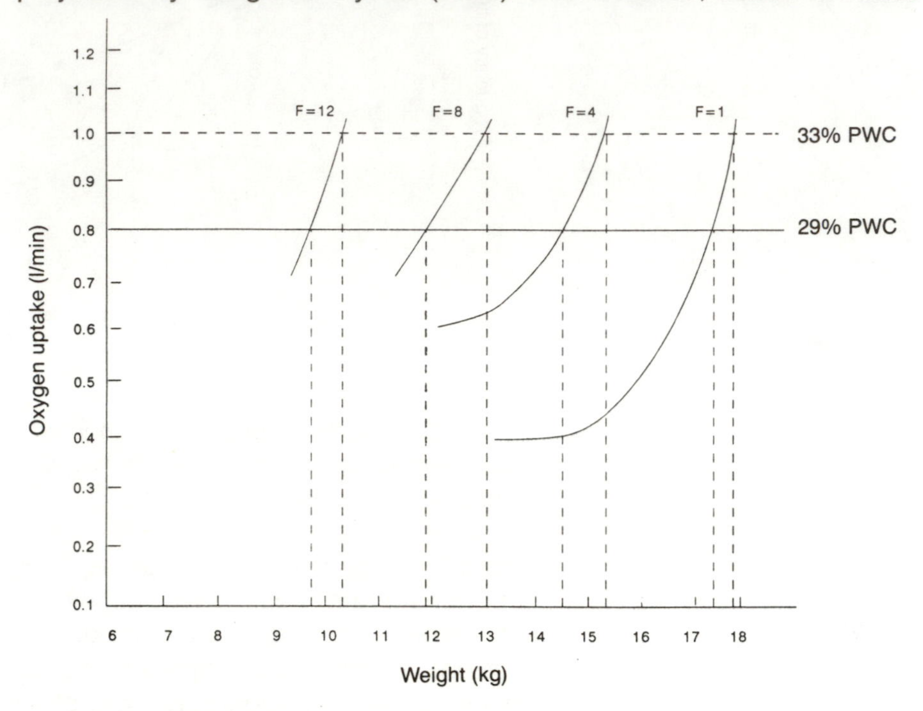

Figure 3.8 Relationship between oxygen uptake, weight, and frequency for industrial males
(Mital, 1985).

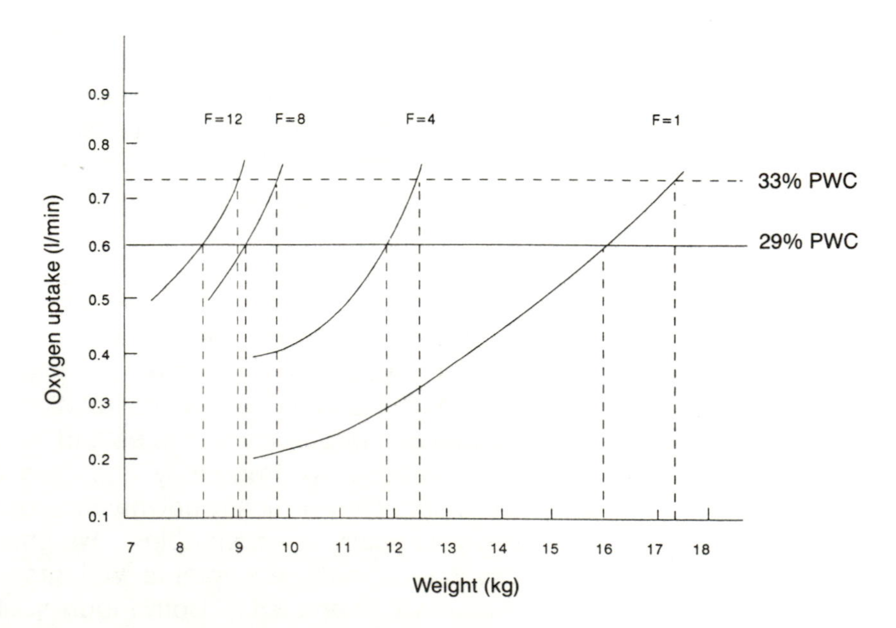

Figure 3.9 Relationship between oxygen uptake, weight, and frequency for industrial females
(Mital, 1985).

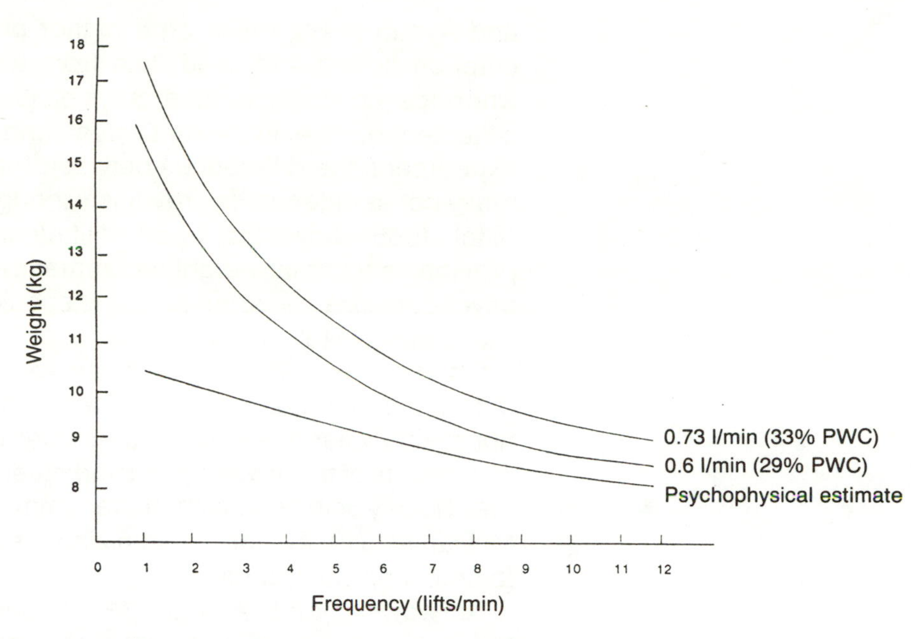

Figure 3.10 Psychophysical and physiological weight estimates for industrial females
(Mital, 1985).

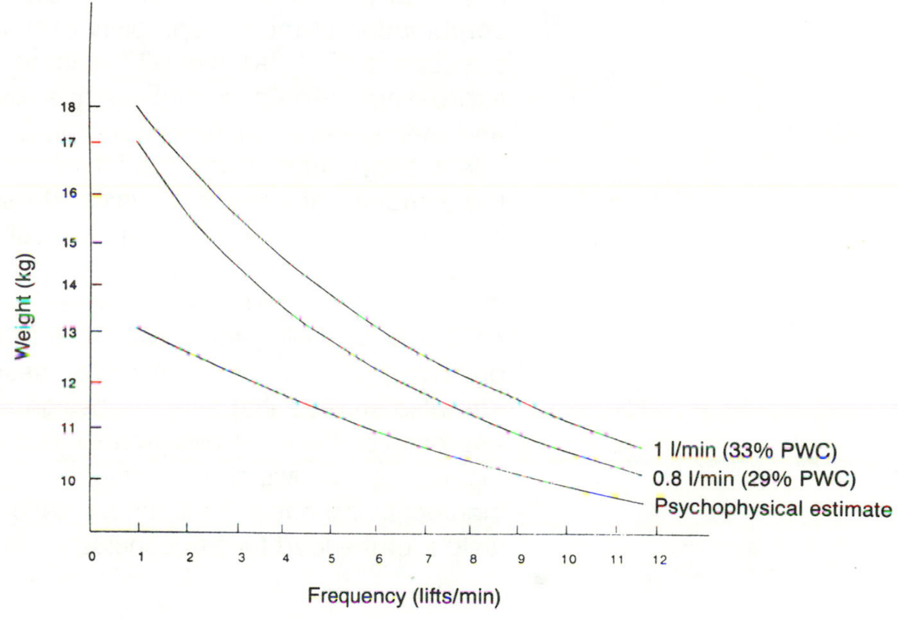

Figure 3.11 Psychophysical and physiological weight estimates for industrial males
(Mital, 1985).

collected from industrial workers, at 1 lift/min, Mital (1985) recommends a weight limit of approximately 13.5 kg for males based on the psychophysical approach; the physiological approach recommendation is close to 17.5 kg. Garg and Ayoub (1980), based on either criterion, however, recommend in excess of 50 kg. At 12 lifts/min, Mital (1985) recommends a weight limit of 9.5–10 kg for males. The recommendation of Garg and Ayoub (1980) varies from 0 kg to approximately 7.5 kg. This recommendation could be interpreted to mean that for frequencies above 9 lifts/min, the physiological approach cannot yield a safe load (Figure 3.7). The experimental data of Mital (1984a, b), however, lead to a very different conclusion. Obviously, these large discrepancies are because Garg and Ayoub (1980) relied on a higher physiological approach design criterion ceiling and used data from a variety of studies. However, when psychophysical and physiological data are simultaneously collected on the same individual (industrial worker) in the same experiment, the differences between the two design approaches are really not as different as previously thought. Furthermore, the study by Mital (1985) shows that above 4 lifts/min the two design approaches provide very similar weight recommendations; at below 4 lifts/min, the psychophysical recommendations should be considered more reliable (Ayoub and Mital, 1989).

Chaffin *et al*. (1983) conducted a series of push/pull experiments to determine the posture individuals would choose (psychophysical approach) when maximizing their exertions with one and two hands. The results of this investigation indicated that subjects chose postures that closely coincided with the recommendations made by Ayoub and McDaniel (1974) for maximizing isometric push/pull exertions (biomechanical approach).

Karwowski and Ayoub (1984) developed fuzzy membership functions for biomechanical, physiological and psychophysical stresses to demonstrate that it was possible to combine biomechanical and physiological lifting stresses to yield the psychophysical lifting stress. The underlying assumption was that a combination of the acceptability of biomechanical and physiological stresses leads to an overall measure of lifting task acceptability as expressed by the psychophysical stress. Based on this assumption, and membership functions generated to demonstrate this synergistic effect, they found conditions for which the acceptability measures of the combined and psychophysical stresses were close together. Thus, Karwowski and Ayoub (1984) were able to demonstrate that it was possible to combine stresses from two different design approaches and equate it to the stress predicted by the third approach should that be the case. Using this concept, Kim (1990) compared biochemical, physiological, and psychophysical design approaches (Figure 3.12). Kim also showed that different design approaches become the limiting approach as the lifting design approach criterion changes. For example, Figure 3.13 shows that for the floor to shoulder height region, the biomechanical design criterion (spinal compression of 7848 N) limits the weight of the load for frequencies up to 1.6 lifts per minute. For higher

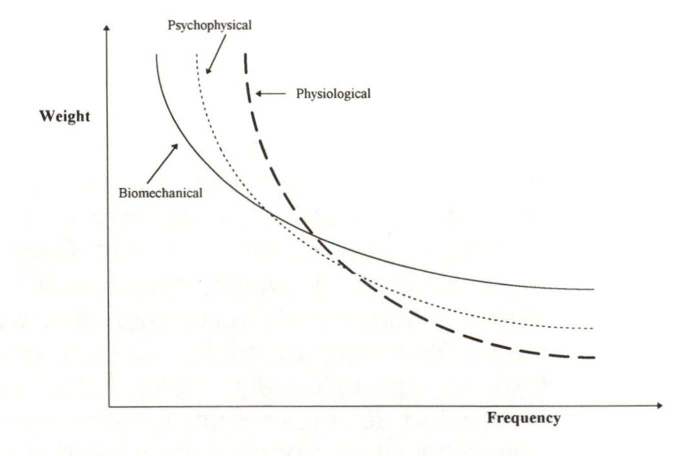

Figure 3.12 Comparison of biomechanical, physiological, and psychophysical design approaches. Modified from Kim (1990)

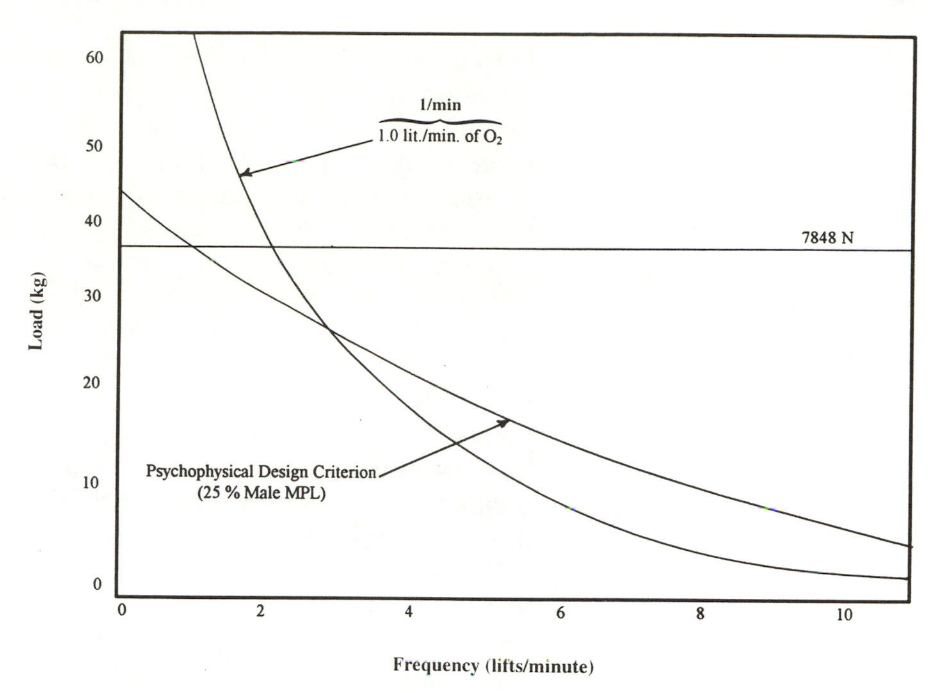

Figure 3.13 Limiting design approach across the frequency range for floor to shoulder height region. Modified from Kim (1990).

frequencies, the weight is first limited by the psychophysical design criterion (25% males for Maximum Permissible Limit, MPL) and then by the physiological design criterion (1 litre of O_2 per minute). The load values and the limiting design approach will obviously be different if the design criteria change.

Nicholson (1989) compared biomechanical (isometric strength and spinal compressive stress, and IAP criteria), psychophysical (maximum acceptable weight of lift criterion), and physiological (metabolic energy expenditure rate criterion) approaches for a variety of MMH activities. The psychophysical data were from Snook (1978), biomechanical stresses were calculated by using the model provided by Chaffin (1975), the IAP data were provided by Davis and Stubbs (1980), and the physiological data were primarily from Petrofsky and Lind (1978) and Legg and Pateman (1985). In order to have an unbiased comparison, comparisons were performed only when experimental conditions were compatible. The conditions considered were MMH activity, population age, percentile ranges, handling frequency and handling posture.

The results of the comparison showed that the weights proposed by IAP data were lower than those based on the psychophysical approach for lower lifting height regions. For higher lifting height regions, the IAP data led to heavier weight recommendations. Psychophysical data also generated compressive forces 10% greater than the NIOSH action limit. However, the psychophysical load recommendations were lower than the biomechanical load recommendations. IAP loads generated compressive forces that were 45 to 60% lower than the action limit. The IAP data and the data by Snook (1978) for pushing showed good agreement. However, for pulling activities the psychophysical force recommendations were lower than the IAP based recommendations.

It should be noted that the current psychophysical load recommendations are substantially lower (Snook and Ciriello, 1991) than those given by Snook (1978). Heavier weight recommendations provided by IAP for higher lifting height regions are also suspect since the muscular strength in these regions is substantially lower than at lower height regions. Furthermore, IAP criteria recommend a single correction factor for frequent handling (30% reduction) regardless of the frequency. In comparing lifting load recommendations, for instance, the same IAP based load recommendations were used for 4 lifts/min as for 9 lifts/min. The psychophysical approach recommends a substantial reduction due to frequency. Given these major differences in the IAP and the data by Snook (1978), it is not surprising that psychophysical load recommendations were greater than IAP load recommendations in some height regions, while the IAP load recommendations were greater in other height regions. This explanation is reinforced by the close agreement observed between the two data sets for infrequent lifting (frequency of less than 1 lift/min).

The review of studies comparing different design approaches appears to leave the impression that the differences between the design approaches are perhaps not as deep as previously believed. Kumar and Mital (1992) compared the margin of safety for the human back provided by the various design approaches. It appears that different design approaches provide a very similar margin of safety for the back.

Margin of safety for the human back

The diverse and seemingly unrelated design approaches have been practised for a long time, often in isolation from each other, and if one were to pursue one logic it may appear to be contradictory to the other. A comprehensive integration of these approaches to arrive at a common recommendation is critical for the safety of workers involved in MMH activities. However, this objective will remain unachievable unless we infer and deduce a margin of safety for the human back.

The margin of safety for the human back, which is influenced by multiple variables as discussed earlier, is defined for simplicity as the difference between the ultimate compressive strength of the lumbar spinal column and the compression endured by the same structure during MMH activities. This definition of the margin of safety does not imply a direct relationship between the physiological, epidemiological and psychophysical design criterion and biomechanical loads. However, applying the same criterion of a difference between the lumbar spinal compression generated in a given activity or a maximum voluntary lift or maximal acceptable effort on the one hand, and the ultimate compressive strength on the other, yields a biomechanical difference. These differences, expressed as percentage of the ultimate compressive strength of a matched sample, have been used as a functional and operational definition of the margin of safety in our discussion in this section. Furthermore, the lumbar spinal compressions reported here for various studies have been obtained from the biomechanical models used in the respective studies.

Epidemiological approach

Snook (1978) proclaimed that up to one-third of back injuries can be prevented. In a survey carried out by the Liberty Mutual Insurance Company, it was found that up to one-quarter of the jobs required manual handling of a magnitude acceptable to less than 75% of the workers. These jobs were responsible for half the back injuries reported, thus indicating a three-fold increase in susceptibility when the tasks require such exertion. This implies that two-thirds of all back injuries associated with MMH activities can be prevented if the job is designed to fit at least 75% of the workforce. The survey extrapolated that one-third of the injuries will occur despite any design effort. Counting the injuries which were not associated with heavy manual handling, Snook (1978) calculated an overall possible reduction in low-back injuries by 33% for industrial workers if an appropriate design strategy were to be instituted. Though such claims are hard to substantiate experimentally, it indicates an approximate 33% collective margin of safety for the workforce.

Ayoub *et al.* (1983) observed a rapid and substantial increase in back problems suffered by male workers as the JSI increased above 1.5. The maximum weight associated with this JSI value was observed to be 27.24 kg. This load provides a spinal compression of 3930 N. When this compression is compared to the mean ultimate spinal compression strength of male spinal elements (5700 N; range approximately 3000–11000 N), provided by Jager and Luttmann (1991), a margin of safety of 31% is obtained. Assuming that females would also show a steep increase in back problems at a JSI value of 1.5, a factor of safety of approximately 31% would provide a spinal compression of 2689 N (mean ultimate compression strength of female spine elements = 3900 N; range approximately 2000–7000 N).

Biomechanical approach

Since pain is a physiological mechanism for ensuring system safety and every back injury is a mechanical perturbation or disturbance of normality associated with such pain episodes, biomechanical aspects become pivotal in such considerations. Conversely, if such biomechanical safety and normality can be maintained for these spinal segments and elements, the problem of low-back pain can be largely controlled. As mentioned in the 'biomechanical approach' section, several researchers have looked into the material properties of these spinal structures. Based on the combined data of Evans and Lissner (1959) and Sonoda (1962), a consensus on the mean ultimate compressive strength of the lumbar spinal column was reached; for people under the age of 40 years, approximately 6700 N, and for people 60 years or over, approximately 3400 N. These values were incorporated by NIOSH in its *Work Practices Guide for Manual Lifting* to provide MPL and AL, respectively (NIOSH, 1981). As generic norms, these values may serve a useful function. However, the concept of safety margin can only be addressed if an individual's maximum voluntary effort, and thereby generated compressive force, is compared with the ultimate compressive strength of their spinal units. Clearly, such a study is impossible for obvious reasons. Therefore, a comparison of voluntarily generated maximal compressive forces of an experimental sample with the ultimate compressive strength of the spinal units of an age-, gender-, and body weight-matched sample may provide a reasonable approach.

Andersson and Schultz (1979) reported a maximal voluntary isometric stoop lifting effort of 900 N from a sample of five males. This was calculated to have generated a compression of 7863 N on the lumbar spine using the model of Schultz and Andersson (1981). Kumar *et al.* (1988) studied isometric and isokinetic lifting strength capabilities at 20, 60 and 100 cm/s linear velocities of load of 10 young males. They reported a mean peak strength of 726 N, 672 N, 639 N, and 597 N, respectively, for these four conditions. Under these experimental conditions, the mean peak compressive forces generated as calculated by the model of Chaffin (1969) were 6933 N, 6329 N, 6017 N and 5613 N, respectively. A compression of 6933 N exceeded the NIOSH maximum permissible limit (MPL) of 6700 N, which supposedly represents the mean ultimate compressive strength of the lumbar spinal column (as determined by Jager and Luttmann (1991), the mean ultimate compressive strength of male spinal elements is 5700 N; for females, this number is 3900 N). Four subjects out of 10 had significantly exceeded the NIOSH MPL. Such efforts are also reported in other studies (Table 3.6). Hutton and Adams (1982) tested the ultimate compressive strength of 16 lumbar spinal columns from eight male subjects, whose age and body weights matched the experimental samples of Andersson and Schultz (1979), Kumar *et al.* (1988), and Kumar and Garand (1991) (Table 3.7).

The mean compressive strength of their sample was 10219 N. Since the age and body weight of the subjects in the five studies were matched, the results may be extrapolated from one study or the other. The significance of body weight lies in the stimulus it provides the skeletal system in establishing mechanical properties. Hutton and

Table 3.6 Maximal voluntary contraction, ultimate compressive strength and margin of safety across selected studies with matched samples (mean values) (Kumar and Mital, 1992)

Study	Lift	Condition	Strength (N)	Compression (N)	Margin of safety (%)
Bartelink (1957)				8920	13
Morris *et al.* (1961)				8869	13
Eie (1966)				6423	37
Andersson and Schultz (1979)	Stoop	Isometric	900	7863	23
Kumar *et al.* (1988)	Stoop	Isometric	726	6933	32
		Isokinetic			
		20 cm/s	672	6329	38
		60 cm/s	639	6017	41
		100 cm/s	597	5613	45
Kumar and Garand (1991)	Stoop	Isometric	700	7813	23
		Isokinetic	580	6186	39
	Squat	Isometic	400	6273	38
		Isokinetic	451	5729	44

Table 3.7 Demographic details of the experimental sample of different studies (Kumar and Mital, 1992)

Study	Gender	*n*	Age range (years)	Weight range (kg)	Height range (cm)
Bartelink (1957)	Male	1	15–58		
Morris *et al.* (1961)	Male				
Eie (1966)	Male				
Andersson and Schultz (1979)	Male	5	22–43		
Hutton and Adams (1982)	Male	8	22–46	65–86	
Kumar *et al.* (1988)	Male	10	23–34	76.5 (Mean)	181 (Mean)
Kumar (1991)	Male	20	18–40	61–87	159–185
Kumar and Garand (1991)	Male	18	21–54	63–80	162–188

n, Sample size.

Adams (1982) have shown a linear relationship between the body weight and the ultimate compressive strength of the spinal unit. The mechanical milieu, therefore, plays an important role in bone growth, density and hypertrophy. This has been further supported by Granhed *et al*. (1987). Though the subjects in the studies by Kumar *et al*. (1988) and Kumar and Garand (1991) were slightly younger, they came from a similar age range. Further, an identical body weight across the three studies ensures similar tissue strength characteristics. A comparison between the mean ultimate compressive strength, as provided by Hutton and Adams (1982), and the magnitude of the compression voluntarily generated reveals a 32% and 23% difference, considering data from Kumar *et al*. (1988) and Andersson and Schultz (1979), respectively (Table 3.6). However, in other studies this difference was found to be as low as 13% (Bartelink, 1957; Morris *et al*., 1961), or as high as 37% (Eie, 1966). Within the same sample, Kumar *et al*. (1988) reported a significant change in the strength capability with a change in the state of the lift. As the isokinetic lift velocity increased, the maximal strength decreased. Comparison of the lumbo-sacral compression, caused as a result of these maximal voluntary contractions, revealed that this margin of safety was least in the static mode of lifting (32%), and increased as the isokinetic lift velocity increased (up to 45%) to 100 cm/s. The validity of this comparison with the data of Hutton and Adams (1982) is further enhanced due to their experimental protocol involving testing of the spinal unit in the flexed position, being the natural posture in which lift is performed.

The structural failure of spinal units can precipitate in either acute or chronic conditions. Most of the studies have concentrated on determination of peak forces in activities of known high stresses. However, biological tissues are like other physical materials with a finite life, and are similarly subject to wear and tear. They are capable of self-repair, but are also subject to mechanical deformation upon load application. All biological tissues are viscoelastic in nature and prolonged load may result in permanent deformation. Repeated load application may also result in cumulative fatigue, reducing their stress-bearing capacity. Such changes may reduce the threshold of stress at which the tissue fails. Kumar (1991) has reported a strong association between cumulative load (biomechanical load and exposure time integral over entire work experience) and low-back pain in an age-, gender-, body weight-, height-, and occupation-matched sample. One may, therefore, examine the fatigue failure characteristics of the spinal column.

Brinckmann *et al*. (1987, 1988) and Hansson *et al*. (1987) investigated the fatigue failure of the lumbar spine. In their experimental protocol, loads of between 20 and 30%, 30 and 40%, 40 and 50%, 50 and 60%, 60 and 70%, and 75% of the estimated ultimate compressive strength of spinal units were applied at a frequency of 0.25 Hz. They found that both the magnitude of the load and the number of cycles affect the spinal column failure. Fatigue failures were produced at lower loads with high repetition and at higher loads with low repetition. When their subjects were loaded at between 50 and 60% of the ultimate compressive strength, 92% of them suffered fatigue failures after 5000 cycles. A 91% fatigue failure rate was reported after 500 cycles when the load was increased by an additional 10%. At a load of 75% of the ultimate compressive strength, the fatigue factors were precipitated in 10 cycles only. Long-range, low-grade grading of the spine will be difficult to control and measure. However, looking at the results of Brinckmann *et al*. (1987, 1988) it would appear that physiological limitations strongly favour biological safety. To begin with, the compression generated by the maximum voluntary contraction ranges between 68 and 77% of the ultimate compressive strength, allowing a 23–32% margin of safety. When one considers that the maximum voluntary contraction (MVC) can be sustained only for a few seconds and that it decays exponentially with the duration of the hold, it is obvious that such compressions cannot be self-generated. Also, the MVC cannot be repeatedly generated without long rest pauses. Rapid repeated trials degenerate quickly with a drastic reduction in magnitude, thereby preventing the total exposure (load × cycle) from rising. In addition, glycogen depletion and lactate accumulation render the human body physiologically unable to accumulate a large number of load cycles due to general exhaustion and muscle fatigue. Therefore, it is stated that the loading frequencies used by these authors are physiologically unattainable even in a highly paced industrial environment. It must also be borne in mind that such rapid cyclic loading does not allow the viscoelastic biological tissues their much-needed recovery time. In turn this will progressively accentuate tissue deformation, rendering them more vulnerable to injury. Even when performing submaximally, Snook (1978) showed that the maximum acceptable load of lift determined psychophysically for males dropped by 30–40% for floor to knuckle and shoulder to arm reach heights, and by around 50% for knuckle to shoulder height lift when the frequency was increased from 1 lift/h to 12 lifts/min. Thus, an initial difference between the ultimate compressive strength and the compression at MVC (25–32%) when combined with the physiological limitation of inability of repetition will only increase the margin of safety over and above the 30% range as stated before.

In addition to failure in compression mode, the tissues can be strained beyond their physiological limit and precipitate injury. To investigate this aspect, Adams and Hutton (1986) compared the maximal *in vivo* range of flexion of the lumbar and lumbo-sacral vertebral joints with that of osteoligamentous preparations. The active range of flexion was reported by them to be 10% short of the osteoligamentous preparation. Such a difference between extreme forward flexion and the elastic limit of the osteoligamentous preparation ensures safety from possible strain injuries by preventing excessive deformation and generation of high tensile stresses. In their experiment, Adams and Hutton (1986) also report that in a typical lumbar motion segment a 2% reduction in flexion at its elastic limit reduces the resistance to bending moment by 50% and reduces the tensile stresses on the intervertebral ligaments and the annulus by 50%. At the limit of the active flexion, the osteoligamentous preparation provides half of the resistance to bending moment exerted by the upper body in forward bending (Adams *et al*., 1980). Considering most activities of daily living and occupational activities, it is obvious that only modest ranges of motion are commonly used. Thus, such an interplay between posture and material properties ensures at least a safety margin of 50% in force enduring capacity. A given degree of muscle contraction is evoked for postural stability and readiness to move to the next phase of activity. Any sudden force may, however, tend to overcome viscoelastic resistance of the muscle due to high strain deformation. Such forces may result in sprains and strains as minor injuries before structural damage can occur.

Kumar and Davis (1983) studied the electromyographical activity (EMG) of erector-spinae muscle and the IAP during a static weight hold and a dynamic lift of the same weight. Both the EMG and IAP values for static posture activities were only one-third to one-half of those obtained for lifting the same weight in dynamic mode. Kumar and Davis (1983) suggested that an increased biomechanical demand in a dynamic lift was due to overcoming inertia and the need for postural stability. Freivalds *et al.* (1984) also found that dynamic effort increased the load on the spine by as much as 40%. Park (1973) reported that acceleration during dynamic lifting increased the biomechanical stress by 15–20%. Since Kumar and Davis (1983) found that the physiological variables for static loading evoked only up to one-third to one-half of the dynamic activity, it is suggested that the excess response was to meet the need of the dynamic lifting activity. Since the acceleration and inertia effects have been shown to increase the load to between 20 and 40% (Park, 1973; Freivalds *et al.*, 1984), the increased response may well contribute from 30 to 35% towards the safety of the execution of the task.

Psychophysical approach

In a simultaneous consideration of psychophysical and physiological approaches for determination of safe load for lifting, Mital (1985) reported that the former always yielded a value less than that provided by the latter across low as well as high frequencies. However, the magnitude of difference was not very large. Mital and Kromodihardjo (1986) reported significant regression between the psychophysically determined maximal acceptable load for lifting and lumbar spinal compressive stresses. The regression equation ($CF = -830 + 17.81 \times PLC$; where CF is the compressive force in newtons and PLC is the psychophysical lifting capacity in newtons) explained 75% of variance in the relationship. Using the data of Hutton and Adams (1982) they extrapolated, by means of this regression equation, the ultimate compressive strength of the lumbar spine of individuals. Comparing the ultimate compressive strength with the compressive load developed during lifting maximal acceptable loads revealed a difference of 30–50% (Table 3.8). This difference was considered to be the margin of safety of the human back. Thus, a self-selection load enhanced and ensured the margin of safety in excess of 30%. However, the sensitivity of psychophysical methodology over different time periods has been the subject of debate (Mital, 1984a, b). The reliability of the physiological sensors feeding the psychophysical

Table 3.8 Margin of safety based on extrapolated strength of spine and peak compressive stress resulting from lifting psychophysically acceptable loads (Mital and Kromodihardjo, 1986)

Body weight (kg)	Compressive force (N)	Spinal strength (N)	Factor of safety (%)
72.8	6478	9300	30
84.8	5574	11100	50
53.5	3489	5350	35
64.4	4400	8000	45

perception over a very long term due to sensory conditioning has not yet been established. It is also unclear if such a sensory feedback may be driven by the long-term safety of body organs. It is conceivable that this methodology may not allow the detection of the threshold level of cumulative load where vulnerability is heightened as shown by Kumar (1991). As a matter of fact, Mital (1987) compared maximum acceptable weight of lift data on 74 inexperienced students with data previously collected for 74 experienced industrial materials handlers under identical experimental conditions to determine the pattern of differences in the two groups. The results revealed that while students and industrial workers did not differ significantly in their physique and isometric strength exertion capabilities, there were very significant differences in the weights the two groups were willing to lift for a regular 8 h shift. Male students accepted 11% less weight on average than industrial males. It may, therefore, be implied that the industrial workers are accumulating the total load exposure by 11% over their work experience. Such an increased exposure may result in reaching the cumulative threshold sooner, thereby precipitating a low-back injury. Sairanen (1980) reported that foresters with 20 years experience had no more complaints of low-back pain than a matched group of sedentary workers, although the foresters had more radiologic signs of disc degeneration. This could imply a sensory conditioning which may tend to increase risk over a long time.

Physiological approach

Though there is no direct biomechanical correlate of the metabolic energy expenditure rate, it may play an important role in precipitation of injury. It may happen by developing tiredness, fatigue and general incoordination, which may result in a biomechanical destabilization of the spinal posture. Unnatural motion under stress may precipitate injury. Therefore, metabolic energy expenditure of MMH activities is an important aspect of safety of the back. As discussed under the 'physiological approach', the design criterion ceiling of 28–29% (bicycle), or 21–23% (uphill treadmill), should serve as the MMH activities design guideline. One may keep the metabolic cost within this guideline, yet maximize the biomechanical load if the frequency was reduced significantly. In such cases, the metabolic cost criterion may be most unsuitable. Furthermore, mechanical fatigue stress on the spinal structure as a result of repeated handling may not be accounted for by the physiological design criterion.

Conclusion

The maximum acceptable weight of handling and maximum voluntary contraction have been calculated to generate a lumbar spinal compression of 50–70% of the ultimate compressive strength of lumbar spinal column. The fatigue failure of the lumbar spinal units can occur by rapid application of maximal voluntary contraction. However, it is physiologically impossible thereby to protect the biomechanical safety of the system. A psychophysical self-selection of workload for lifting activity has also been found to favour system safety by arriving at a load equal to or lower than arrived at by using physiological criteria. It, therefore, appears that psychophysical criteria may have an integrative role. A comparative representation of the safety margin by various design approaches is presented in Table 3.9.

Table 3.9 Margin of safety with different approaches based on the ultimate compressive strength reported by Hutton and Adams (1982) (From Kumar and Mital, 1992)

Design approach[a]	Margin of safety (%)
Biomechanical	12–45
Psychophysical	30–45
EMG	33–50
IAP	50

[a]Epidemiological approach provides a margin of safety of 31% when load compression is compared against the mean ultimate compression strength of spinal column elements provided by Jager and Luttmann (1991).

References

Adams, M.A. and Hutton, W.C., 1982. Prolapsed intervertebral disc: a hyperflexion injury. *Spine*, **7**, 184–191.

Adams, M.A. and Hutton, W.C., 1985. The effect of posture on the lumbar spine. *Journal of Bone and Joint Surgery*, **67B**, 625–629.

Adams, M.A. and Hutton, W.C., 1986. Has the lumbar spine a margin of safety in forward bending? *Clinical Biomechanics*, **1**, 3–6.

Adams, M.A., Hutton, W.C. and Stott, J.R.R.R., 1980. The resistance to flexion of the lumbar intervertebral joint. *Spine*, **5**, 245–253.

Andersson, G.B.J., 1981. Epidemiologic aspects of low-back pain in industry. *Spine*, **6**, 53–59.

Andersson, G.B.J. and Schultz, A.B., 1979. Transmission of moments across the elbow joint and the lumbar spine. *Journal of Biomechanics*, **12**, 747–755.

Andersson, G.B.J., Ortengren, R. and Nachemson, A., 1977. Intradiskal pressure, intra-abdominal pressure and myoelectric back muscle activity related to posture and loading. *Clinical Orthopaedics and Related Research*, **129**, 156–164.

Aquilano, N.J., 1968. A physiological evaluation of time standards for strenuous work as set by stopwatch time study and two predetermined motion time data systems. *Journal of Industrial Engineering*, **19**, 425–432.

Arad, D. and Ryan, M.D., 1986. The incidence and prevalence in nurses of low back pain: a definitive survey exposes the hazards. *Australian Nurses Journal*, **16**, 44–48.

Astrand, I., 1967. Aerobic work capacity. *Circulation Research*, **20**, **21** (Suppl. I), 211–217.

Astrand, P.O. and Rodahl, K., 1986. *Textbook of Work Physiology*, 3rd edn. New York: McGraw-Hill.

Ayoub, M.M. and McDaniel, J.W., 1974. Effect of operator stance on pushing and pulling tasks. *Transactions of the American Institute of Industrial Engineers*, **6**, 185–195.

Ayoub, M.M. and Mital, A., 1989. *Manual Materials Handling*. London: Taylor & Francis.

Ayoub, M.M., Bethea, N.J., Bobo, M., Burford, C.L., Caddel, K., Intaranont, K., Morrissey, S. and Selan, J.L., 1981. *Mining in Low Coal, Vol. 1, Biomechanics and Work Physiology*. Final Report, US Bureau of Mines Contract No. H03087022.

Ayoub, M.M., Gidcumb, C.F., Hafez, H., Intaranont, K., Jiang, B.C. and Selan, J.L., 1983. *A Design Guide for Manual Lifting Tasks*. Prepared for OSHA.

Bartlink, D.L., 1957. The role of abdominal pressure in relieving the pressure on the lumbar intervertebral discs. *Journal of Bone and Joint Surgery*, **39B**, 718–725.

Bergquist-Ullman, M. and Larson, U., 1977. Acute low back pain in industry. *Acta Orthopedica Scandinavica*, Suppl 170, 1–117.

Biggemann, M., Hilweg, D. and Brinckmann, P., 1988. Prediction of the compressive strength of vertebral bodies of the lumbar spine by quantitative computed tomography. *Skeletal Radiology*, **17**, 264–269.

Bink, B., 1962. The physical working capacity in relation to working time and age. *Ergonomics*, **5**, 25–28.

Brinckmann, P., Johannelweling, N., Hilweg, D. and Biggemann, M., 1987. Fatigue fracture of human lumbar vertebrae. *Clinical Biomechanics*, **2**, 94–97.

Brinckmann, P., Biggemann, M. and Hilweg, D., 1988. Fatigue fractures of human lumbar vertebrae. *Clinical Biomechanics*, **3** (Suppl 1).

Brinckmann, P., Biggemann, M. and Hilweg, D., 1989. Prediction of the compressive strength of human lumbar vertebrae. *Clinical Biomechanics*, **4** (Suppl 2) 1–13.

Brown, T., Hansen, R. and Yorra, A., 1957. Some mechanical tests on the lumbosacral spine with particular reference to the intervertebral discs: a preliminary report. *Journal of Bone and Joint Surgery*, **39A**, 1135–1164.

Chaffin, D.B., 1969. A computerized biomechanical model: development of and use in studying gross body actions. *Journal of Biomechanics*, **2**, 429–442.

Chaffin, D.B., 1975. On the validity of biomechanical models of the low back for weight lifting analysis. *Proceedings of the American Society of Mechanical Engineers*, 75-WA-Bio-1, pp. 1–13. New York: American Society of Mechanical Engineers.

Chaffin, D.B. and Park, K.S., 1973. A longitudinal study of low-back pain as associated with occupational weight lifting factors. *American Industrial Hygiene Association Journal*, **32**, 513–525.

Chaffin, D.B., Herrin, G.D., Keyserling, W.M. and Foulke, J.A., 1976. *Pre-employment Strength Testing in Selecting Workers for Materials Handling Jobs*. Cincinnati: National Institute for Occupational Safety and Health. Publication CDC-99-74-62.

Chaffin, D.B., Herrin, G.D. and Keyserling, W.M., 1978. Preemployment strength testing: an updated position. *Journal of Occupational Medicine*, **20**,

403–408.

Chaffin, D.B., Andres, R.O. and Garg, A., 1983. Volitional postures during maximal push/pull exertions in the sagittal plane. *Human Factors*, **25**, 541–550.

Ciriello, V.M. and Snook, S.H., 1983. A study of size, distance, height, and frequency effects on manual handling tasks. *Human Factors*, **25**, 473–483.

Ciriello, V.M., Snook, S.H., Blick, A.C. and Wilkinson, P.L., 1990. The effects of task duration on psychophysically-determined maximum acceptable weights and forces. *Ergonomics*, **33**, 187–200.

Damkot, D.K., Pope, M.H., Lord, J. and Frymoyer, J.W., 1984. The relationship between work history, work environment and low-back pain in men. *Spine*, **9**, 395–399.

Davis, P.R., 1985. Intratruncal pressure mechanisms. *Ergonomics*, **28**, 293–297.

Davis, P.R. and Stubbs, D.A., 1978. Safe levels of manual forces for young males. *Applied Ergonomics*, **9**, 33–37.

Davis, P.R. and Stubbs, D.A., 1980. *Force Limits in Manual Work*. Guildford: IPC Science and Technology Press.

Davis, P.R., Troup, J.D.G. and Burnard, J.H., 1965. Movements of the thoracic and lumbar spine when lifting: a chronocyclographic study. *Journal of Anatomy*, **99**, 13–26.

Durnin, J.V.G.A. and Passmore, R., 1967. *Energy, Work and Leisure*. London: Heinemann Educational.

Eie, N., 1966. Load capacity of the low back. *Journal of Oslo City Hospital*, **16**, 73–98.

Evans, F.G. and Lissner, H.R., 1959. Biomechanical studies on the lumbar spine and pelvis. *Journal of Bone and Joint Surgery*, **41A**, 278–290.

Farfan, H.F., Cossette, J., Robertson, G., Wells, R. and Kraus, H., 1970. Effects of torsion on the intervertebral joint: the role of torsion in the production of disc degeneration. *Journal of Bone and Joint Surgery*, **52A**, 468.

Fernandez, J.E., 1985. Psychophysical lifting capacity over extended periods. PhD dissertation, Texas Tech University, Lubbock, Texas, USA.

Freivalds, A., Chaffin, D.B., Garg, A and Lee, K.S., 1984. A dynamic biomechanical evaluation of lifting maximum acceptable loads. *Journal of Biomechanics*, **17**, 251–262.

Friedman, G.D., 1974. *Primer of Epidemiology*. New York: McGraw-Hill.

Frymoyer, J.W., Pope, M.H., Costanza, M.C., Rosen, J.C., Goggin, J.E. and Wilder, D.G., 1980. Epidemiologic studies of low-back pain. *Spine*, **5**, 419–423.

Frymoyer, J.W., Pope, M.H., Clements, J.H., Wilder, D.G., McPherson, B., Ashikaga, T. and Vermont, B., 1983. Risk factors in low-back pain. *Journal of Bone and Joint Surgery*, **65A**, 213–218.

Garg, A. and Ayoub, M.M., 1980. What criteria exist for determining how much load can be lifted safely? *Human Factors*, **22**, 475–486.

Garg, A. and Saxena, U., 1982. Maximum frequency acceptable to female workers for one-handed lifts in the horizontal plane. *Ergonomics*, **25**, 839–853.

Glover, J.R., 1970. Occupational health research and the problem of back pain. *Transactions of the Society of Occupational Medicine*, **21**, 2–12.

Granhed, H., Jonson, R. and Hansson, T., 1987. The loads on the lumbar spine during extreme weight lifting. *Spine*, **12**, 146–149.

Hamilton, B.J. and Chase, R.B., 1969. A work physiology study of the relative effects of pace and weight in a carton handling task. *Transactions of the American Institute of Industrial Engineers*, **1**, 106–111.

Hansson, T.H., Ross, B. and Nachemson, A., 1980. The bone mineral content and ultimate compressive strength of lumbar vertebrae. *Spine*, **5**, 46–55.

Hansson, T.H., Keller, T.S. and Spengler, D.M., 1987. Mechanical behavior of the human lumbar spine II. Fatigue strength during dynamic compressive loading. *Journal of Orthopaedic Research*, **5**, 479–487.

Herrin, G.D., Jariedi, M. and Anderson, C.K., 1986. Prediction of overexertion injuries using biomechanical and psychophysical models. *American Industrial Hygiene Association Journal*, **47**, 322–330.

Hutton, W.C. and Adams, M.A., 1982. Can the lumbar spine be crushed in heavy lifting? *Spine*, **7**, 586–590.

Hutton, W.C., Cyron, B.M. and Stott, J.R.R., 1979. The compressive strength of lumbar vertebrae. *Journal of Anatomy*, **129**, 753–758.

Jager, M. and Luttmann, A., 1991. Compressive strength of lumbar spine elements related to age, gender, and other influencing factors. In *Electromyographical Kinesiology*, edited by P.A. Anderson, D.J. Hobart and J.V. Danoff, pp. 291–294. Amsterdam: Elsevier Science.

Jensen, R.C., 1988. Epidemiology of work-related back pain. *Topics in Acute Care and Trauma Rehabilitation*, **2**, 1–15.

Karwowski, W. and Ayoub, M.M., 1984. Fuzzy modelling of stresses in manual lifting tasks. *Ergonomics*, **27**, 641–649.

Kelsey, J.L., Githens, P.B., White III, A.A., Holford, T.R., Walter, S.D., O'Conor, T., Ostfeld, A.M., Weil, U., Southwick, W.O. and Calogero, J.A., 1984. An epidemiologic study of lifting and twisting on the job and risk for acute prolapsed lumbar intervertebral disc. *Journal of Orthopaedic Research*, **2**, 61–66.

Khalil, T.M., Genaidy, A.M., Asfour, S.S. and Vinciguerra, T., 1985. Physiological limits in lifting. *American Industrial Hygiene Association Journal*, **46**, 220–224.

Kim, H.-K., 1990. Development of a model for combined ergonomic approaches in manual materials handling tasks. PhD dissertation, Texas Tech University, Lubbock, Texas, USA.

Kromodihardjo, S. and Mital, A., 1986. Kinetic analysis of manual lifting activities: Part I – Development of a three dimensional computer model. *International Journal of Industrial Ergonomics*, **1**, 77–90.

Kromodihardjo, S. and Mital, A., 1987. Biomechanical analysis of manual lifting tasks. *Journal of Biomechanical Engineering*, **109**, 132–138.

Kumar, S., 1980. Physiological responses to weight lifting in different planes. *Ergonomics*, **23**, 987–993.

Kumar, S., 1984. The physiological cost of three different methods of lifting in sagittal and lateral planes. *Ergonomics*, **27**, 425–433.

Kumar, S., 1991. Cumulative load as a risk factor for low-back pain. *Spine*, **15**, 1311–1316.

Kumar, S. and Davis, P.R., 1983. Spinal loading in static and dynamic postures: EMG and intra-abdominal pressure study. *Ergonomics*, **26**, 913–922.

Kumar, S. and Garand, D., 1991. Static and dynamic lifting strength at different reach distances in symmetrical and asymmetrical planes. *Technical Report*, Department of Physical Therapy, University of Alberta, Edmonton, Canada.

Kumar, S. and Mital, A., 1992. Margin of safety for the human back: a probable consensus based on published studies. *Ergonomics*, **35**, 769–781.

Kumar, S., Chaffin, D.B. and Redfern, M., 1988. Isometric and isokinetic back and arm lifting strengths:

device and measurement. *Journal of Biomechanics*, **21**, 35–44.

Lawrence, J.S., 1955. Rheumatism in coal miners, Part III: Occupational factors. *British Journal of Industrial Medicine*, **12**, 249–261.

Legg, S.J., 1981. The effect of abdominal muscle fatigue and training on the intra-abdominal pressure developed during lifting. *Ergonomics*, **24**, 191–195.

Legg, S.J. and Myles, W.S., 1985. Metabolic and cardiovascular cost, and perceived effort over an 8 hour day when lifting loads selected by the psychophysical method. *Ergonomics*, **28**, 337–343.

Legg, S.J. and Pateman, C.M., 1985. Human capabilities in repetitive lifting. *Ergonomics*, **28**, 309–321.

Leskinen, T.P.J., Stalhammer, H.R., Kuorinka, I.A.A. and Troup, J.D.G., 1983a. A dynamic analysis of spinal compression with different lifting techniques. *Ergonomics*, **26**, 595–604.

Leskinen, T.P.J., Stalhamamer, H.R., Kuorinka, I.A.A. and Troup, J.D.G., 1983b. The effect of inertia factors on spinal stress when lifting. *Engineering in Medicine*, **12**, 87–89.

Lin, H.S., Liu, Y.K. and Adams, K.H., 1978. Mechanical response of the lumbar intervertebral joint under complex (physiological) loading. *Journal of Bone and Joint Surgery*, **60-A**, 41–55.

McBroom, R.J., Hayes, W.C., Edwards, W.T., Goldberg, R.P. and White III, A.A., 1985. Prediction of vertebral body compressive fracture using quantitative computed tomography. *Journal of Bone and Joint Surgery*, **67-A**, 1206–1214.

McGill, S.M. and Norman, R.W., 1985. Dynamically and statically determined low back moments during lifting. *Journal of Biomechanics*, **18**, 877–885.

Mairiaux, P.H., Davis, P.R., Stubbs, D.A. and Baty, D., 1984. Relation between intra-abdominal pressure and lumbar moments when lifting weights in the erect posture. *Ergonomics*, **27**, 883–894.

Michael, E.D., Hutton, K.E. and Horvath, S.M., 1961. Cardiorespiratory responses during prolonged exercise. *Journal of Applied Physiology*, **16**, 997–999.

Mital, A., 1983. The psychophysical approach in manual lifting – a verification study. *Human Factors*, **25**, 485–491.

Mital, A., 1984a. Comprehensive maximum acceptable weight of lift database for regular 8-hour workshifts. *Ergonomics*, **27**, 1127–1138.

Mital, A., 1984b. Maximum weights of lift acceptable to male and female industrial workers for extended workshifts. *Ergonomics*, **27**, 1115–1126.

Mital, A., 1985. A comparison between psychophysical and physiological approaches across low and high frequency ranges. *Journal of Human Ergology*, **14**, 59–64.

Mital, A., 1987. Patterns of differences between the maximum weights of lift acceptable to experienced and inexperienced materials handlers. *Ergonomics*, **30**, 1137–1147.

Mital, A., 1992. Psychophysical capacity of industrial workers for lifting symmetrical and asymmetrical loads symmetrically and asymmetrically for 8-hour work shifts. *Ergonomics*, **35**, 745–754.

Mital, A. and Asfour, S.S., 1983. Maximum frequency acceptable to males for one-handed horizontal lifting in the sagittal plane. *Human Factors*, **25**, 563–571.

Mital, A. and Kromodihardjo, S., 1986. Kinetic analysis of manual lifting activities: Part II – Biomechanical analysis of task variables. *International Journal of Industrial Ergonomics*, **1**, 91–101.

Mital, A., Fard, H.F., Khaledi, H. and Channaveeraiah, C., 1987a. Are manual lifting weight limits based on the physiological approach realistic and practical? In *Trends in Ergonomics/Human Factors IV*, edited by S.S. Asfour, pp. 973–977. Amsterdam: North-Holland.

Mital, A., Karwowski, W. and Chalaka, A., 1987b. A laboratory simulation of self-paced and machine-paced industrial stacking and palletizing tasks. *Journal of Human Ergology*, **16**, 31–41.

Morris, J.M., Lucas, D.B. and Bresler, B., 1961. Role of the trunk in stability of the spine. *Journal of Bone and Joint Surgery*, **43-A**, 327–351.

Muller, E.A., 1953. Physiological basis of rest pauses in heavy work. *Quarterly Journal of Experimental Physiology*, **38**, 205–215.

Nachemson, A.L., 1982. The natural course of low back pain. In *Proceedings of the American Academy of Orthopaedic Surgeons Symposium on Idiopathic Low Back Pain*, edited by A.A. White and S.L. Gordson, pp. 46–51. St. Louis: C.V. Mosby.

Nachemson, A., Andersson, G.B.J. and Schultz, A.B., 1986. Valsalva maneuver biomechanics: effects of lumbar trunk loads on elevated intra-abdominal pressure. *Spine*, **11**, 476–479.

National Institute for Occupational Safety and Health, 1981. *Work Practices Guide for Manual Lifting*. Publication No. 81–122. Cincinnati: NIOSH.

Nicholson, A.S., 1989. A comparative study of methods for establishing load handling capabilities. *Ergonomics*, **32**, 1125–1144.

Park, K., 1973. A computerized simulation model of postures during manual materials handling. PhD dissertation, University of Michigan, Ann Arbor, Michigan.

Perey, O., 1957. Fracture of the vertebral end plate in the lumbar spine: an experimental biomechanical investigation. *Acta Orthopedica Scandinavica*, Suppl XXV, 1–101.

Petrofsky, J.S. and Lind, A.R., 1978. Comparison of metabolic and ventilatory responses of men to various lifting tasks and bicycle ergometry. *Journal of Applied Physiology: Respiratory, Environmental and Exercise Physiology*, **45**, 60–63.

Porter, R.W., Adams, M.A. and Hutton, W.C., 1989. Physical activity and the strength of lumbar spine. *Spine*, **14**, 201–203.

Rowe, M.L., 1983. *Backache at Work*. Fairport, New York: Perinton Press.

Sairanen, E., 1980. Felling work, low back pain, and osteoarthritis. *Scandinavian Journal of Work, Environment and Health*, **7**, 18–30.

Schultz, A.B. and Andersson, G.B.J., 1981. Analysis of loads on lumbar spine. *Spine*, **6**, 76–82.

Schultz, A.B., Andersson, G.B.J., Ortengren, R., Haderspeck, K. and Nachemson, A., 1982. Loads on the lumbar spine. *Journal of Bone and Joint Surgery*, **64-A**, 713–720.

Sharp, M.A., Harman, E., Vogel, J.A., Knapik, J.J. and Legg, S.J., 1988. Maximal aerobic capacity for repetitive lifting: comparison with three standard exercise testing modes. *European Journal of Applied Physiology*, **57**, 753–760.

Snook, S.H., 1978. The design of manual handling tasks. *Ergonomics*, **21**, 963–985.

Snook, S.H. and Ciriello, V.M., 1974. Maximum weights and work loads acceptable to female workers. *Journal of Occupational Medicine*, **16**, 527–534.

Snook, S.H. and Ciriello, V.M., 1991. The design of manual handling tasks: revised tables of maximum acceptable weights and forces. *Ergonomics*, **34**, 1197–1213.

Snook, S.H. and Irvine, C.H., 1968. Maximum frequency of lift acceptable to male industrial workers.

American Industrial Hygiene Association Journal, **29**, 531–536.

Snook, S.H., Campanelli, R.A. and Hart, J.W., 1978. A study of three preventive approaches to low back injury. *Journal of Occupational Medicine*, **20**, 478–481.

Sonoda, T., 1962. Studies on the compression, tension, and torsion strength of the human vertebral column. *Journal of Kyoto Prefectural Medical University*, **71**, 659–702.

Takala, E.P. and Kukkonen, R., 1987. The handling of patients on geriatric wards: a challenge for on-the-job training. *Applied Ergonomics*, **18**, 17–22.

Thomson, K.D., 1988. On the bending movement capability of the pressurized abdominal cavity during human lifting activities. *Ergonomics*, **31**, 817–828.

Troup, J.D.G., Martin, J.W. and Lloyd, C.E.F., 1981. Back pain in industry: a prospective survey. *Spine*, **6**, 61–69.

Yamada, H., 1970. In *Strength of Biological Materials*, edited by F.G. Evans. Baltimore: Williams & Wilkins.

Part II

Chapter 4

Lifting

Two-handed lifting

Over the years, several different design approaches have been used to develop a design database for two-handed symmetrical lifting by industrial workers. These design approaches are reviewed in Chapter 3 of this guide. As our discussion showed, the various design approaches use different design criteria. The diversity in design approaches and design criteria provide different safe load recommendations for manual lifting. As also stated in Chapter 2, a number of individual, task and environmental factors influence an individual's capability to lift loads. For a design database to be realistic and applicable to a wide variety of situations, it should not only account for the influence of these factors, it *must not* violate any of the design criteria.

Earlier discussion also indicates that no design approach can provide safe load recommendations across the entire range of the most significant task factor – namely the lifting frequency. While the biomechanical design approach, since it ignores the effect of repetition, would be most suitable for infrequent load lifting, the physiological design approach would appear to be the logical choice at high lifting frequencies. The psychophysical design approach has been used for low as well as high lifting frequencies, and there is some evidence to indicate that it may overestimate the lifting capabilities of individuals both at very low and very high lifting frequencies.

In this section, we present manual lifting design data for two-handed lifting for both male and female industrial workers for a wide range of task conditions. These task conditions are determined by the various task variables that define the work being performed.

Since the psychophysical design approach provides design data across almost the entire lifting frequency range, we began with that database and modified it in regions where either the biomechanical design criterion or the physiological design criterion was the limiting factor.

The design criteria

The design criteria used in the development of a comprehensive database are based on all four design approaches. **For the epidemiological design approach, the design criterion was the JSI value of 1.5. The maximum load that could be handled for this JSI value was 27.24 kg.**

The biomechanical criterion was the lumbar spine compression that, on average, provided a margin of safety of at least 30% for the lower back. This value was 3930 N for males and corresponded to a load of 27.24 kg. For females, this value was 2689 N and corresponded to a load of approximately 20 kg.

The physiological design criterion was an energy expenditure rate of 4 kcal/min for males and 3 kcal/min for females for a working duration of 8 h. These values reflect approximately 29% and 28% of the physical work capacity (bicycle) of males and

females, respectively. For shorter or longer duration, these values are modified as per Table 4.1.

Table 4.1 Values of the physiological design criterion as a function of working duration (energy expenditure rate in kcal/min and percentage physical work capacity, given in parenthesis)

	Duration (h)			
Gender	1	4	8	12
Male	4.7 (34)[a]	4.4 (32)	4 (29)	3.2 (23)
Female	3.4 (32)[a]	3.2 (30)	3 (28)	2.6 (24)

[a]Bicycle.

Design database

Tables 4.2 and 4.3 provide the design database for male and female industrial workers. The regions where either the biomechanical design criterion or the physiological design criterion is the limiting criterion are highlighted. The data in Tables 4.2 and 4.3 are for two-handed symmetrical lifting of a symmetrical load for a working duration of 8 h and are a function of the following task-related factors: lifting frequency, lifting height, and box-size (expressed in terms of the distance of the load dimension in the sagittal plane; dimension of the load away from the body). The data are given for the 10th, 25th, 50th, 75th and 90th population percentile; the user may choose the population percentile that is to be protected on the job being designed.

These basic data must be modified to reflect the working population's capacity for realistic lifting tasks. The task and environmental factors that may be used to modify the basic design data in Tables 4.2 and 4.3 are:

1. Working duration
2. Limited headroom (spatial restraint)
3. Asymmetrical lifting
4. Load asymmetry
5. Couplings
6. Load placement clearance
7. Heat stress

Part I of this guide discusses these factors and their influence on the MMH capabilities of workers. The correction factors for modifying the basic design data, in order to account for the effect of the above factors, are given in Tables 4.4 to 4.10. A limited discussion on the extent of influence these task and environmental factors have on the manual lifting capability of individuals follows.

Working duration multiplier

The manual lifting capability of workers is significantly influenced by the lifting duration. For shorter durations (less than 8 h), as Table 4.1 shows, more metabolic energy is available and individuals can lift heavier loads; the load is limited by the biomechanical design criterion. For longer durations (more than 8 h), the fatigue buildup requires the weight of the load to be reduced. In general, the lifting capability of industrial males declined by 3.4% for every hour increase in the working duration. The corresponding number for females is 2%.

Table 4.2 Recommended weight of lift (kg) for male industrial workers for two-handed symmetrical lifting for 8 h.

		Frequency of Lift							
Box-size (cm)	Percentile	1/8 h	1/30 min	1/5 min	1/min	4/min	8/min	12/min	16/min
				Floor to 80 cm height					
75	90	17	14	14	11	9	***7***	6	4.5
	75	24	21	20	16	13	***10.5***	9	7
	50	27[a]	27[a]	27	22	17	***14***	12	9.5
	25	27[a]	27[a]	27[a]	27[a]	21	***17.5***	15	12
	10	27[a]	27[a]	27[a]	27[a]	25	***20.5***	18	14.5
49	90	20	17	16	13	10	***7***	7	6.5
	75	27[a]	24	24	19	14	***10***	***10***	9
	50	27[a]	27[a]	27[a]	26	19	***15***	***12.5***	***10***
	25	27[a]	27[a]	27[a]	27[a]	24	18.5	***15***	***12***
	10	27[a]	27[a]	27[a]	27[a]	27[a]	***22***	***17.5***	***15***
34	90	23	19	19	15	11	***7***	***7***	6.5
	75	27[a]	27[a]	27[a]	22	17	***10***	***10***	9.5
	50	27[a]	27[a]	27[a]	27[a]	22	***15***	***14***	***12***
	25	27[a]	27[a]	27[a]	27[a]	27[a]	***20***	***17***	***14***
	10	27[a]	27[a]	27[a]	27[a]	27[a]	***25***	***21***	***15***
				Floor to 132 cm height					
75	90	15	13	13	10	8	***6***	6	4
	75	22	20	19	14.5	12	***10***	9	7
	50	27[a]	25	24	20	15	13	11	9
	25	27[a]	27[a]	27[a]	24.5	18	15	12	11
	10	27[a]	27[a]	27[a]	27[a]	22	19	16	13
49	90	18	16	15	12.5	9	***6***	***6***	***5***
	75	27	22.5	22.5	18	14	***10***	***9***	***8***
	50	27[a]	27[a]	27[a]	24	18	***14***	***12***	***10***
	25	27[a]	27[a]	27[a]	27[a]	22	18	***14***	***11***
	10	27[a]	27[a]	27[a]	27[a]	27	21	***17***	***14***
34	90	22	18	18	14	11	***6***	***6***	***5***
	75	27[a]	26	25	21	16	***10***	***9***	***8***
	50	27[a]	27[a]	27[a]	27[a]	22	***14***	***12***	***10***
	25	27[a]	27[a]	27[a]	27[a]	27	***20***	***14***	***11***
	10	27[a]	27[a]	27[a]	27[a]	27[a]	21	***17***	***14***
				Floor to 183 cm height					
75	90	15	12	12	9.5	8	***6***	5	3
	75	21	18	17	14	11	***9***	8	6
	50	27[a]	24	23	19	15	12	10	8
	25	27[a]	27[a]	27[a]	24	18	14	12	9
	10	27[a]	27[a]	27[a]	27[a]	22	18	15	12
49	90	17	15	14	11	9	***6***	***6***	***4***
	75	24	21	21	16	12	***9***	***9***	***7***
	50	27[a]	27[a]	27[a]	22	16	***14***	***12***	***10***
	25	27[a]	27[a]	27[a]	27[a]	20	17	***14***	***11***
	10	27[a]	27[a]	27[a]	27[a]	23	20	***17***	***14***
34	90	20	16	16	13	9	***6***	***6***	***4***
	75	27[a]	24	24	19	15	***9***	***9***	***7***
	50	27[a]	27[a]	27[a]	26	19	***14***	***12***	***10***
	25	27[a]	27[a]	27[a]	27[a]	23	***20***	***14***	***11***
	10	27[a]	27[a]	27[a]	27[a]	27[a]	24	***17***	***14***

Table 4.2 Recommended weight of lift (kg) for male industrial workers for two-handed symmetrical lifting for 8 h. (Continued)

		Frequency of Lift							
Box-size (cm)	Percentile	1/8 h	1/30 min	1/5 min	1/min	4/min	8/min	12/min	16/min
		80 cm to 132 cm height							
75	90	19	18	16	15	13	***7***	***6***	***5***
	75	25	23	21	20	17	***8***	***8***	***7***
	50	27[a]	27[a]	26	25	21	***12***	***11***	***9***
	25	27[a]	27[a]	27[a]	27[a]	26	***17***	***13***	***12***
	10	27[a]	27[a]	27[a]	27[a]	27[a]	23	20	***16***
49	90	19	18	16	15	13	***7***	***6***	***5***
	75	25	23	21	20	17	***8***	***8***	***7***
	50	27[a]	27[a]	26	25	21	***12***	***11***	***9***
	25	27[a]	27[a]	27[a]	27[a]	26	***17***	***13***	***12***
	10	27[a]	27[a]	27[a]	27[a]	27[a]	23	20	***16***
34	90	22	20	18	17	14	***7***	***6***	***5***
	75	27[a]	26	23	22	18	***8***	***8***	***7***
	50	27[a]	27[a]	27[a]	27[a]	23	***12***	***11***	***9***
	25	27[a]	27[a]	27[a]	27[a]	27	***17***	***13***	***12***
	10	27[a]	27[a]	27[a]	27[a]	27[a]	24	21	***16***
		80 cm to 183 cm height							
75	90	16	15	13	12	11	***7***	***6***	***5***
	75	22	20	18	17	15	***8***	***8***	6
	50	27[a]	25	23	21	19	***12***	***11***	8
	25	27[a]	27[a]	27	26	23	***17***	***13***	11
	10	27[a]	27[a]	27[a]	27[a]	27	22	18	13
49	90	16	15	13	12	11	***7***	***6***	***5***
	75	22	20	18	17	15	***8***	***8***	6
	50	27[a]	25	23	21	19	***12***	***11***	8
	25	27[a]	27[a]	27	26	23	***17***	***13***	11
	10	27[a]	27[a]	27[a]	27[a]	27	22	18	13
34	90	18	17	15	14	12	***7***	***6***	***5***
	75	24	22	20	19	16	***8***	***8***	7
	50	27[a]	27[a]	25	24	20	***12***	***11***	9
	25	27[a]	27[a]	27[a]	27[a]	24	20	16	12
	10	27[a]	27[a]	27[a]	27[a]	27[a]	22	18	13
		132 cm to 183 cm height							
75	90	15	14	12	12	9	***7***	6	4
	75	20	18	15	15	12	***9***	8	6
	50	25	23	20	19	16	***12***	10	7
	25	27[a]	27	25	23	19	***15***	***12***	10
	10	27[a]	27[a]	27[a]	27	22	***17***	***13***	12
49	90	18	16	14	14	11	***7***	7	5
	75	23	21	19	18	14	***9***	***8***	6
	50	27[a]	27	24	23	18	***12***	***10***	9
	25	27[a]	27[a]	27[a]	27[a]	21	***15***	***12***	10
	10	27[a]	27[a]	27[a]	27[a]	25	***17***	***13***	11
34	90	20	18	17	16	13	***7***	***6***	5
	75	26	24	22	21	17	***9***	***8***	8
	50	27[a]	27[a]	27[a]	26	21	***12***	***11***	10
	25	27[a]	27[a]	27[a]	27[a]	25	***15***	***14***	13
	10	27[a]	27[a]	27[a]	27[a]	27[a]	***17***	***16***	15

[a]Weight limited by the biomechanical design criterion (3930 N spinal compression). Numbers in bold italics, weight limited by the physiological design criterion (4 kcal/min).

Table 4.3 Recommended weight of lift (kg) for female industrial workers for two-handed symmetrical lifting for 8 h.

Box-size (cm)	Percentile	Frequency of Lift 1/8 h	1/30 min	1/5 min	1/min	4/min	8/min	12/min	16/min
Floor to 80 cm height									
75	90	12	9	8	7	7	6	5	4
	75	14	11	10	9	9	8	7	6
	50	17	13	12	11	10	9	8	7
	25	20[a]	15	14	13	12	11	9	7
	10	20[a]	17	16	14	14	13	11	9
49	90	13	9	8	8	8	7	6	5
	75	16	12	10	10	9	8	7	6
	50	19	14	13	12	11	10	9	8
	25	20[a]	17	15	14	13	11	10	8
	10	20[a]	19	17	15	15	13	11	9
34	90	15	11	10	9	9	8	7	6
	75	19	14	13	12	11	9	8	7
	50	20[a]	17	16	14	13	11	10	8
	25	20[a]	20[a]	18	17	15	13	12	10
	10	20[a]	20[a]	20[a]	19	18	15	13	11
Floor to 132 cm height									
75	90	10	7.5	6.5	6	6	5	4	3
	75	12	9	8	7.5	7.5	6.5	6	5
	50	14	11	10	9	8	7.5	6.5	6
	25	17	12.5	11.5	11	10	9	7.5	6.5
	10	19	14	13	11.5	11.5	11	9	8
49	90	11	7.5	6.5	6.5	6.5	6	5	4
	75	13	10	8	8	7.5	6.5	6	5
	50	16	11.5	11	10	9	8	7.5	6.5
	25	17	14	12.5	11.5	11	9.5	8	7
	10	19	16	14	12.5	12.5	11	9	7.5
34	90	12.5	9	8	7.5	7.5	6.5	6	5
	75	16	11.5	11	10	9	8	6.5	5.5
	50	19	14	13	11.5	11	9.5	8	7
	25	20[a]	17	15	14	12.5	11	10	9
	10	20[a]	19	17	16	15	13	11	9
Floor to 183 cm height									
75	90	9	6	6	5	5	4.5	4	3
	75	11	8	7	7	7	6	5	4.5
	50	12.5	10	9	8	7	7	6	5.5
	25	15	11	10	10	9	8	7	6
	10	17	12.5	12	10	10	10	8	7
49	90	10	7	6	6	6	5.5	4.5	3.5
	75	12	9	7	7	7	6	5	4.5
	50	14	10	10	9	8	7	7	6
	25	15	12	11	10	10	8.5	7	6.5
	10	17	14	12	11	11	10	8	7
34	90	11	8	7	7	7	6	5	4.5
	75	14	10	10	9	8	7	6	5
	50	17	12	12	10	10	8.5	7	6
	25	20	15	13.5	12	11	10	9	8
	10	20[a]	17	15	14	13.5	12	10	8

Table 4.3 Recommended weight of lift (kg) for female industrial workers for two-handed symmetrical lifting for 8 h. (Continued)

		Frequency of Lift							
Box-size (cm)	Percentile	1/8 h	1/30 min	1/5 min	1/min	4/min	8/min	12/min	16/min
				80 cm to 132 cm height					
75	90	13	11	10	9	8	***6***	6	5
	75	15	13	12	11	9	***7***	7	6
	50	17	15	14	13	11	***9***	9	8
	25	20	17	16	14	12	11	10	9
	10	20[a]	19	17	16	14	12.5	11	9.5
49	90	13	11	10	9	8	***6***	6	5
	75	15	13	12	11	9	***7***	7	6
	50	17	15	14	13	11	***9***	9	8
	25	20	17	16	14	12	11	10	9
	10	20[a]	19	17	16	14	12.5	11	9.5
34	90	14	12	11	10	9	***7***	***6.5***	***6.5***
	75	17	14	13	12	11	***8.5***	***8.5***	8
	50	19	17	15	14	13	***11***	10	8.5
	25	20[a]	19	17	16	14	13.5	***11.5***	11
	10	20[a]	20[a]	19	18	16	14.5	13	11.5
				80 cm to 183 cm height					
75	90	11	9.5	9	8	7	***5***	5	4.5
	75	13	11	10.5	9.5	8	***6***	6	5
	50	15	13	12	11	10	***8***	8	7
	25	17.5	15	14	12	10.5	10	9	8
	10	19	17	15	14	12	11	10	8
49	90	11	9.5	9	8	7	***5***	5	4.5
	75	13	11	10.5	9.5	8	***6***	6	5
	50	15	13	12	11	10	***8***	8	7
	25	17.5	15	14	12	10.5	10	9	8
	10	19	17	15	14	12	11	10	8
34	90	12	10.5	10	9	8	***6***	***6***	***6***
	75	15	12	11	10.5	10	***7.5***	***7.5***	7
	50	17	15	13	12	11	***10***	9	7.5
	25	19	17	15	14	12	11	***10***	10
	10	20[a]	19	17	16	14	13	11	10
				132 cm to 183 cm height					
75	90	9	8	7	7	7	***5***	***4***	***3***
	75	11	9	9	8	8	***6***	***5***	***4***
	50	13	11	10	9	9	8	7	6
	25	14	12	11	10	10	9	8	7
	10	16	14	13	12	11	10	9	8
49	90	10	9	8	7	7	***5***	***4***	***3***
	75	12	10	9	9	8	***6***	***5***	***4***
	50	14	12	11	10	9	8	7	6
	25	15	13	12	11	10	9	8	7
	10	17	15	14	13	11	10	9	8
34	90	12	11	10	9	8	***6***	***6***	***6***
	75	14	12	11	11	9	***7***	***7***	***7***
	50	17	14	13	12	11	***9***	9	8
	25	19	16	15	14	12	11	10	9
	10	20[a]	18	16	15	14	***12***	11	9.5

[a]Weight limited by the biomechanical design criterion (2689 N spinal compression). Numbers in bold italics, weight limited by the physiological design criterion (3 kcal/min).

Table 4.4 Working duration (h) multiplier[a]

Gender	Duration (h) 1	4	8	12
Men	1.238	1.136	1.000	0.864
Women	1.140	1.080	1.000	0.920

[a]Interpolate for intermediate or neighbouring working duration.

Table 4.4 shows the multipliers that the data in Tables 4.2 and 4.3 should be adjusted to reflect the workers' lifting capability for the actual lifting duration (note that data for one lift every 8 h, in Tables 4.2 and 4.3, can only be adjusted for durations longer than 8 h). **It should be noted that the upper weight limit for males and females is 27 kg and 20 kg, respectively. Therefore, if after working duration adjustment, the recommended weight obtained is more than 27 kg for males, or 20 kg for females, it must be substituted by these limits.** For instance, the lifting capability of 10th percentile males for the floor to 80 cm height for an 8 h working duration for a frequency of 8 lifts/min and a box-size of 34 cm is 25 kg (from Table 4.2). If this task is to be performed only for 1 h, the duration multiplier from Table 4.4 will be 1.238. This would change the weight recommendation to 30.95 kg (25 kg × 1.238). However, since the maximum weight is limited by the biomechanical design criterion, 30.95 kg weight recommendation for 1 h working duration must be replaced with 27 kg weight recommendation.

Limited headroom multiplier

The data in Tables 4.2 and 4.3 are for a situation that permits workers to stand fully upright. Sometimes it is not possible to stand up straight due to limited headroom (low ceilings). The inability to stand upright influences manual lifting capacity. Table 4.5 provides multipliers to adjust the data in Tables 4.2 and 4.3 in situations that do not allow upright posture.

Table 4.5 Limited headroom multiplier

Stature[a]	Fully upright	95% upright	90% upright	85% upright	80% upright
Multiplier[a]	1.00	0.60	0.40	0.38	0.36

[a]Interpolate for intermediate stature.

Asymmetrical lifting multiplier

Asymmetrical lifting by 90° can reduce a person's lifting capability by 8.4% if the feet move, and by as much as 22% if the feet are not allowed to move. Since in most working situations feet will move (starting and ending locations of the load will change), the effect of asymmetrical lifting will not be as severe as when the feet are kept fixed. However, in order to account for both situations, the asymmetrical lifting multipliers given in Table 4.6 should be used. **To minimize spinal rotation, and hence the shear stress on the spine, it is recommended that workers move their feet when turning with the load.**

Load asymmetry multiplier

Most loads are not symmetrical. The design data in Tables 4.2 and 4.3 takes into consideration load c.g. shift away from the body (in the

Table 4.6 Asymmetrical lifting multiplier

Angle of turn (deg)[a]	Multiplier[a]
0–30	1.0000
30–60	0.924[b]
60–90	0.848[b]
Above 90	0.800[c]

[a]Interpolate for intermediate values.
[b]These corrections may be too high if feet are allowed to move.
[c]Assuming people will move their feet.

sagittal plane). Many loads, however, have their c.g. shifted sideways (in the frontal plane). The sideway shift of load c.g. also affects manual lifting capability of individuals. The multiplier for load asymmetry (c.g. shift sideways) is given in Table 4.7.

Table 4.7 Load asymmetry (c.g. shift sideways in the frontal plane) multiplier

Load asymmetry[a] (cm)	Multiplier[a]
0	1.00
10	0.96
20	0.89
30	0.84

[a]Interpolate for intermediate or neighbouring values.

Couplings multiplier

Absence of handles or good coupling also affects manual lifting capabilities. The reduction in capability could be as much as 15% if the load has no handles or holds to initiate the lift. The basic design data in Tables 4.2 and 4.3 must be adjusted by the coupling multiplier given in Table 4.8.

Table 4.8 Couplings multiplier

Couplings	Multiplier
Good and comfortable handles/firm holds to initiate the lift	1.000
Poor quality handles/limited or slippery hold	0.925
No handles/holds to initiate the lift	0.850

Load placement clearance multiplier

Frequently loads are placed in narrow locations, such as shelves with limited clearance. This requires loads to be manoeuvered carefully to their final destination before being released. Careful manoeuvering increases the load holding time and reduces the lifting capability. Data in Tables 4.2 and 4.3 should be adjusted by the load placement clearance (distance between the load and the obstruction, e.g. shelf wall or another load) multiplier, given in Table 4.9, in the event loads are to be placed in narrow spaces with limited side clearance.

Table 4.9 Load placement clearance multiplier[a]

Load clearance (mm)	Multiplier
Unlimited to 30	1.00
15	0.91
3	0.87

[a]Interpolate for intermediate or neighbouring values.

Heat stress multiplier

The basic design data given in Tables 4.2 and 4.3 were generated under no heat stress conditions. Frequently, lifting is performed under hot and humid climate. The additional physical stress caused by the heat stress results in reduced manual lifting capabilities. Therefore, if the work is to be performed in hot and humid climate, a heat stress multiplier must be used to adjust the basic design data. The value of the multiplier is given in Table 4.10.

Table 4.10 Heat stress (WBGT) multiplier[a]

Heat stress (WBGT)	Multiplier
Up to 27°C	1.00
At 32°C	0.88

[a]Interpolate for intermediate values.

Use of tables

In order to determine the lifting capability of a specific population, the following relationship should be used:

A = lifting capability of a specific population percentile for a given lifting height, lifting frequency, and box-sixe

A = B × C

where

B = lifting capability of the population percentile as given by Table 4.2 or Table 4.3 for the specified lifting height, lifting frequency and box-size

C = working duration multiplier × limited headroom multiplier × asymmetrical lifting multiplier × load asymmetry multiplier × couplings multiplier × load placement clearance multiplier × heat stress multiplier

One-handed lifting

One-handed lifting is a relatively infrequent activity in the workplace. Even when lifting is performed with only one hand, the load is generally moved in the horizontal plane; rarely is one involved in moving loads vertically with one hand. For these reasons, very little design data are available for one-handed lifting. The data that are available are in two forms: (1) maximum frequencies that can be sustained for moving fixed loads in the horizontal plane back and forth

Table 4.11 Maximum frequencies (cycles/min) acceptable to males for one-handed horizontal lifting

			Percentile				
Posture	Reach (cm)	Load (kg)	90	75	50	25	10
Sitting	38.1	2.27	6	11	15	18	24
		4.54	6	8	10	12	14
		6.81	5	6	7	8	8
	63.5	2.27	6	10	14	18	22
		4.54	6	8	10	12	14
		6.81	5	6	7	8	8
Standing	38.1	2.27	7	11	15	20	23
		4.54	7	9	11	13	14
		6.81	5	6	7	8	9
	63.5	2.27	6	10	13	16	21
		4.54	6	8	11	13	14
		6.81	5	6	7	7	8

in the sagittal plane and (2) the maximum weights that can be lifted infrequently in the vertical plane. When using these design data for design purposes, it should be noted that these data are for the stronger hand and **we recommend that whenever loads are lifted using only one hand, the stronger hand be used.**

Table 4.11 present the maximum frequencies (lifting cycles/min) for one-handed horizontal lifting of various loads in the sagittal plane acceptable to male workers for a 2-h working period. The lifting cycle is defined as: reach for the load (Figure 4.1, position A), pick up the load, move the load away from the body (position A to B), release the load, and return to the starting point without the load. The maximum one-handed lifting frequencies acceptable to female workers for a 2-h period are given in Table 4.12.

Table 4.12 Maximum frequencies (cycles/min) acceptable to females for one-handed horizontal lifting

			Percentile				
Posture	Reach (cm)	Load (kg)	90	75	50	25	10
Sitting	38.1	2.27	8	9	10	12	13
		4.54	7	7	7	8	8
	63.5	2.27	7	7	8	8	9
		4.54	4	5	5	5	6
Standing	38.1	2.27	8	9	10	11	11
		4.54	5	5	6	7	8
	63.5	2.27	7	8	8	10	11
		4.54	4	5	5	7	8

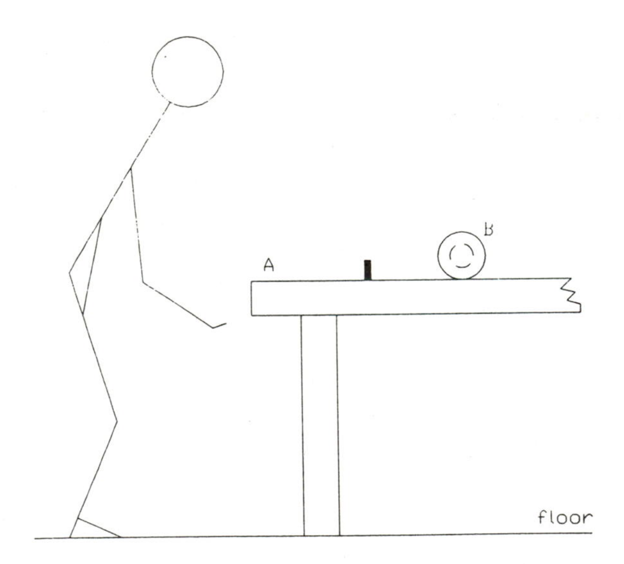

Figure 4.1 One-handed lifting in the sagittal plane.

The maximum weights that can be lifted up vertically infrequently have been determined for males using the IAP biomechanical design criterion. For the standing posture, the maximum weight recommendations vary from 9 to 30 kg, depending upon the age of the worker and how close the load is to the shoulder. It should be noted that weight recommendations above 27 kg will violate the spinal compression criterion. Furthermore, when the load is handled dynamically, the distance between the load and shoulder changes. It is not uncommon to move a load from close to the body to a location away from the body. In such situations, the distance between the shoulder and the load will increase continuously and, therefore, weight must be reduced. **Given how the loads are generally moved in the vertical plane, we recommend that the maximum weight that males can lift infrequently with their stronger hand while standing should not exceed 9 kg. For the sitting posture, the maximum load that males should be permitted to lift infrequently should not exceed 8 kg. For females, the maximum permissible weight for one-handed infrequent lifting should not exceed 6 kg.**

Two-person lifting

Frequently, the load being lifted is large and awkwardly shaped and cannot be lifted by one person. In such cases, it is desirable to lift the load using two people. Few studies have looked at two-person team lifting. Some of these studies indicate that in a team, individuals are able to lift significantly more weight than the sum of their individual capacities. Other studies have shown that team lifting capacity is approximately 10% lower than the sum of the individual members' capacities. This is particularly true when the individuals have to squat.

Since the information available in the published literature is conflicting, the team lifting capacity may be determined by doubling the value from Tables 4.2 and 4.3. This value should be adjusted for the various factors as discussed in the two-handed lifting section. The maximum weight of the load that a two-person male team should lift should not exceed 54 kg. A two-person female team should not lift a load weighing more than 40 kg. Furthermore, the members of the team should be of similar stature.

Chapter 5

Pushing

Unlike lifting, where the load is completely supported by the worker, pushing is a material handling activity that requires objects to be pushed on floors. In this case, the load is supported by the floor and force is exerted to move the object. Thus, pushing activities are not as

Table 5.1 Recommended initial forces (kg) for male (female) industrial workers for two-handed pushing

		Frequency of pushing[a]							
Handle height (cm)	Population percentile	10/min	5/min	4/min	2.4/min	1/min	1/2 min	1/5 min	1/8 h
				2.1 m pushing distance					
144 (135)	90	20 (14)	22 (15)			25 (17)		26 (20)	31 (22)
	75	26 (17)	29 (18)			32 (21)		34 (24)	41 (27)
	50	32 (20)	36 (22)			40 (25)		42 (29)	51 (32)
	25	38 (24)	43 (25)			47 (29)		50 (33)	61 (37)
	10	44 (26)	49 (28)			55 (33)		58 (38)	70 (41)
95 (89)	90	21 (14)	24 (15)			26 (17)		28 (20)	34 (22)
	75	28 (17)	31 (18)			34 (21)		36 (24)	44 (27)
	50	34 (20)	38 (22)			43 (25)		45 (29)	54 (32)
	25	41 (24)	46 (25)			51 (29)		54 (33)	65 (37)
	10	47 (26)	53 (28)			59 (33)		62 (38)	75 (41)
64 (57)	90	19 (11)	22 (12)			24 (14)		25 (16)	31 (18)
	75	25 (14)	28 (15)			31 (17)		33 (19)	40 (21)
	50	31 (16)	35 (17)			39 (20)		41 (23)	50 (25)
	25	38 (19)	42 (20)			46 (23)		49 (27)	59 (30)
	10	43 (21)	48 (23)			53 (26)		57 (30)	68 (33)
				7.6 m pushing distance					
144 (135)	90			14 (15)		21 (16)		22 (18)	26 (20)
	75			18 (18)		27 (19)		28 (22)	34 (24)
	50			23 (21)		33 (23)		35 (26)	42 (29)
	25			27 (25)		40 (27)		42 (31)	51 (34)
	10			31 (28)		46 (30)		48 (34)	58 (38)
95 (89)	90			16 (14)		23 (16)		25 (19)	30 (21)
	75			21 (17)		30 (20)		32 (22)	39 (25)
	50			26 (20)		38 (23)		40 (27)	48 (30)
	25			31 (23)		45 (27)		48 (31)	58 (34)
	10			35 (26)		52 (31)		55 (35)	66 (39)
64 (57)	90			13 (11)		20 (14)		21 (16)	26 (17)
	75			16 (14)		26 (17)		27 (19)	33 (21)
	50			20 (16)		32 (20)		34 (23)	41 (25)
	25			25 (19)		39 (23)		41 (27)	50 (29)
	10			28 (22)		45 (26)		47 (30)	57 (33)
				15.2 m pushing distance					
144 (135)	90				16 (12)	19 (14)		20 (15)	25 (17)
	75				21 (15)	25 (17)		26 (19)	32 (21)
	50				26 (18)	31 (20)		33 (22)	40 (25)
	25				31 (20)	37 (23)		40 (26)	48 (29)
	10				36 (23)	43 (26)		45 (29)	55 (32)
95 (89)	90				18 (11)	22 (14)		23 (16)	28 (17)
	75				24 (14)	28 (17)		30 (19)	36 (21)
	50				29 (16)	35 (20)		37 (23)	45 (25)
	25				35 (19)	42 (23)		45 (27)	54 (29)
	10				40 (22)	49 (26)		52 (30)	62 (33)
64 (57)	90				15 (9)	19 (12)		20 (13)	24 (15)
	75				19 (11)	24 (14)		26 (16)	31 (18)
	50				23 (14)	30 (17)		32 (19)	39 (21)
	25				28 (16)	36 (20)		39 (23)	47 (25)
	10				32 (18)	42 (22)		44 (25)	54 (28)

Table 5.1 Recommended initial forces (kg) for male (female) industrial workers for two-handed pushing (Continued)

		Frequency of pushing[a]							
Handle height (cm)	Population percentile	10/min	5/min	4/min	2.4/min	1/min	1/2 min	1/5 min	1/8 h
					30.5 m pushing distance				
144 (135)	90					15 (12)		19 (14)	24 (17)
	75					19 (15)		25 (17)	31 (21)
	50					24 (18)		31 (21)	38 (25)
	25					28 (20)		37 (24)	46 (29)
	10					32 (23)		42 (27)	53 (33)
95 (89)	90					17 (12)		22 (15)	27 (18)
	75					21 (15)		28 (18)	35 (21)
	50					27 (18)		35 (21)	44 (26)
	25					32 (21)		42 (24)	52 (30)
	10					37 (24)		48 (28)	60 (33)
64 (57)	90					14 (11)		19 (12)	23 (15)
	75					18 (13)		24 (15)	30 (18)
	50					23 (15)		30 (18)	37 (22)
	25					28 (18)		36 (21)	45 (25)
	10					32 (20)		41 (23)	52 (28)
					45.7 m pushing distance				
144 (135)	90					13 (12)		16 (14)	20 (17)
	75					16 (15)		21 (17)	26 (21)
	50					20 (18)		26 (21)	33 (25)
	25					24 (20)		32 (24)	39 (29)
	10					28 (23)		36 (27)	45 (33)
95 (89)	90					14 (12)		19 (15)	23 (18)
	75					18 (15)		24 (18)	30 (21)
	50					23 (18)		30 (21)	37 (26)
	25					27 (21)		36 (24)	45 (30)
	10					32 (24)		41 (28)	52 (33)
64 (57)	90					12 (11)		16 (12)	20 (15)
	75					16 (13)		21 (15)	26 (18)
	50					20 (15)		26 (18)	32 (22)
	25					24 (18)		31 (21)	39 (25)
	10					27 (20)		36 (23)	44 (28)
					61 m pushing distance				
144 (135)	90						12 (12)	14 (13)	18 (15)
	75						16 (14)	18 (15)	23 (19)
	50						20 (17)	22 (18)	28 (22)
	25						23 (20)	27 (21)	34 (26)
	10						27 (22)	31 (24)	39 (29)
95 (89)	90						14 (12)	16 (13)	20 (16)
	75						18 (15)	21 (16)	26 (19)
	50						22 (18)	26 (19)	32 (23)
	25						27 (20)	31 (22)	38 (27)
	10						31 (23)	35 (25)	44 (30)
64 (57)	90						12 (10)	14 (11)	17 (13)
	75						15 (12)	18 (13)	22 (16)
	50						19 (15)	22 (16)	28 (19)
	25						23 (17)	26 (19)	33 (23)
	10						26 (19)	30 (21)	38 (25)

[a]Interpolate for intermediate frequencies. Values in parentheses are for females.

strenuous as manual lifting. However, the prevalence of these activities in industry requires that these activities be designed to fit the human capability.

Two-handed pushing

Generally, two-handed pushing force data have been expressed in three forms: (1) maximum dynamic force that can be exerted to set an object in motion (initial force), (2) maximum dynamic force that can keep an object in motion (sustained force), and (3) maximum isometric force that can be exerted while trying to push an object. The last category represents the absolute maximum force that workers can exert. Since the biomechanical design criterion is not the limiting factor in a frequent pushing task, we started with the psychophysical design data and adjusted it such that the physiological design criterion was not violated. Table 5.1 provides the maximum initial pushing force database for males and females based on the psychophysical and physiological design criteria. The maximum sustained pushing force database for males and females based on the psychophysical and physiological criteria are given in Table 5.2.

The maximum isometric pushing force data for males and females are given in Figure 5.1 The figure also provides optimum foot distance and handle height, as a percentage of reach height, that maximize isometric pushing force.

One-handed pushing

The maximum pushing force for one-handed horizontal pushing is based on the IAP design criterion. It should be noted that when objects are pushed, the arms are neither fully extended nor completely flexed; a shoulder-grip distance equal to half the arm length is more realistic.

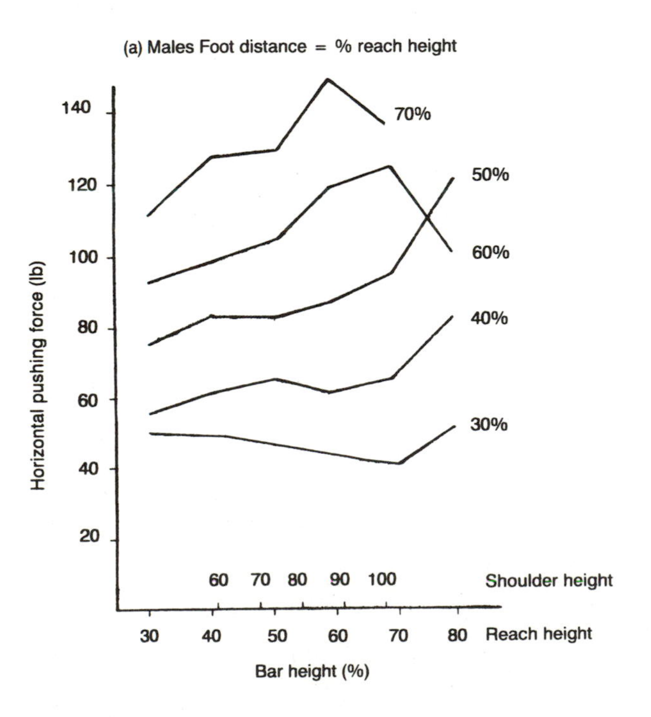

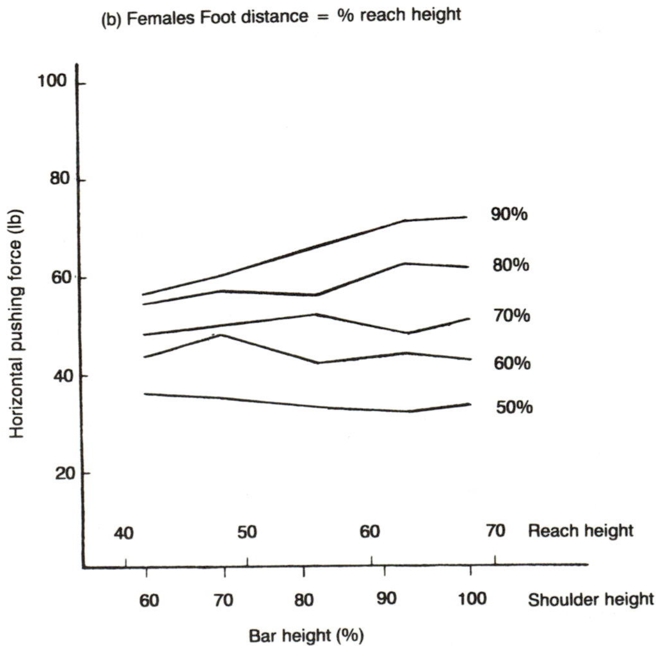

Figure 5.1 Isometric pushing force (kg) for different foot distances and handle heights.

Table 5.2 Recommended sustained forces (kg) for male (female) industrial workers for two-handed pushing

		Frequency of pushing[a]							
Handle height (cm)	Population percentile	10/min	5/min	4/min	2.4/min	1/min	1/2 min	1/5 min	1/8 h
				2.1 m pushing distance					
144 (135)	90	10 ***(5)***	13 (8)			15 (10)		18 (11)	22 (14)
	75	***10 (8)***	17 ***(10)***			21 (14)		24 (16)	30 (21)
	50	***13 (10)***	***21 (14)***			27 (19)		31 (21)	38 (28)
	25	***17 (14)***	***21 (17)***			33 (24)		38 (27)	47 (36)
	10	***20 (15)***	***25 (20)***			38 (28)		45 (32)	54 (42)
95 (89)	90	10 ***(5)***	13 (7)			16 (9)		19 (10)	23 (13)
	75	***11 (7)***	18 ***(9)***			22 (13)		25 (15)	31 (19)
	50	***14 (9)***	***18 (13)***			28 (18)		33 (20)	40 (26)
	25	***17 (12)***	***22 (15)***			34 (22)		40 (25)	49 (33)
	10	***21 (14)***	***26 (19)***			40 (26)		46 (30)	57 (39)
65 (57)	90	10 ***(4)***	13 (6)			16 (8)		18 (9)	23 (12)
	75	***11 (6)***	18 ***(8)***			21 (11)		25 (13)	31 (17)
	50	***14 (8)***	***18 (11)***			28 (15)		32 (17)	39 (23)
	25	***17 (10)***	***22 (14)***			34 (19)		39 (22)	48 (29)
	10	***21 (13)***	***26 (16)***			39 (23)		46 (26)	56 (34)
				7.6 m pushing distance					
144 (135)	90			***6 (5)***		13 (7)		15 (8)	18 (11)
	75			***8 (7)***		17 (11)		20 (12)	25 (16)
	50			***10 (10)***		22 ***(10)***		26 (16)	32 (21)
	25			***13 (12)***		28 ***(15)***		32 (20)	39 (27)
	10			***15 (15)***		***26 (18)***		38 (24)	46 (32)
95 (89)	90			***6 (5)***		13 (8)		15 (9)	18 (11)
	75			***9 (7)***		17 (11)		20 (13)	25 (17)
	50			***11 (10)***		22 ***(13)***		26 (17)	32 (22)
	25			***14 (12)***		27 ***(16)***		32 (21)	39 (28)
	10			***16 (14)***		***26 (19)***		37 (25)	45 (33)
65 (57)	90			***6 (5)***		12 (7)		14 (8)	18 (11)
	75			***9 (7)***		17 (10)		19 (12)	24 (15)
	50			***11 (9)***		21 ***(12)***		25 (16)	31 (21)
	25			***14 (12)***		26 ***(15)***		31 (20)	37 (26)
	10			***16 (14)***		***24 (18)***		36 (23)	44 (31)
				15.2 m pushing distance					
144 (135)	90				***6 (4)***	11 ***(4)***		13 (7)	16 (9)
	75				***9 (6)***	15 ***(8)***		18 (10)	22 (13)
	50				***11 (9)***	***16 (10)***		23 (14)	28 (18)
	25				***13 (10)***	***19 (13)***		28 (17)	34 (22)
	10				***16 (12)***	***22 (15)***		33 (20)	40 (27)
95 (89)	90				***6 (4)***	11 ***(4)***		13 (7)	16 (10)
	75				***9 (6)***	15 ***(8)***		18 (11)	21 (14)
	50				***11 (8)***	***15 (11)***		23 (14)	28 (19)
	25				***14 (10)***	***19 (14)***		28 (18)	34 (24)
	10				***16 (12)***	***22 (16)***		32 (21)	40 (28)
64 (57)	90				***6 (4)***	11 ***(4)***		12 (7)	15 (9)
	75				***9 (6)***	14 ***(8)***		17 (10)	21 (13)
	50				***11 (8)***	***15 (10)***		22 (13)	27 (17)
	25				***14 (10)***	***18 (13)***		27 (17)	33 (22)
	10				***17 (12)***	***21 (15)***		31 (20)	38 (26)

Table 5.2 Recommended sustained forces (kg) for male (female) industrial workers for two-handed pushing (Continued)

		Frequency of pushing[a]							
Handle height (cm)	Population percentile	10/min	5/min	4/min	2.4/min	1/min	1/2 min	1/5 min	1/8 h
				30.5 m pushing distance					
144 (135)	90					***6 (4)***		12 (6)	16 (8)
	75					***9 (6)***		16 (9)	21 (12)
	50					***12 (9)***		20 (12)	28 (16)
	25					***14 (11)***		25 ***(13)***	34 (21)
	10					***17 (13)***		29 ***(14)***	39 (25)
95 (89)	90					***6 (4)***		12 (6)	16 (9)
	75					***9 (7)***		16 (9)	21 (13)
	50					***12 (9)***		20 (12)	27 (17)
	25					***14 (11)***		25 ***(13)***	33 (22)
	10					***17 (14)***		29 ***(15)***	39 (26)
64 (57)	90					***6 (4)***		11 (6)	15 (8)
	75					***9 (6)***		15 (8)	20 (12)
	50					***11 (9)***		19 (11)	26 (16)
	25					***14 (10)***		24 (14)	32 (20)
	10					***16 (13)***		28 ***(14)***	37 (24)
				45.7 m pushing distance					
144 (135)	90					***5 (4)***		10 (5)	13 (8)
	75					***8 (6)***		13 (8)	18 (11)
	50					***9 (8)***		17 ***(9)***	23 (15)
	25					***12 (9)***		21 ***(11)***	28 (19)
	10					***15 (12)***		24 ***(14)***	33 (22)
95 (89)	90					***5 (4)***		9 (6)	13 (8)
	75					***7 (6)***		13 (8)	18 (12)
	50					***9 (8)***		17 ***(9)***	23 (16)
	25					***12 (10)***		21 ***(12)***	28 (20)
	10					***14 (12)***		24 ***(15)***	32 (24)
64 (57)	90					***5 (4)***		9 (5)	13 (7)
	75					***7 (6)***		12 (8)	17 (11)
	50					***9 (8)***		16 ***(9)***	22 (15)
	25					***11 (9)***		20 ***(11)***	27 (19)
	10					***14 (11)***		23 ***(14)***	31 (22)
				61 m pushing distance					
144 (135)	90						7 ***(3)***	8 (4)	11 (6)
	75						***7 (4)***	11 (6)	15 (9)
	50						***9 (7)***	14 ***(7)***	19 (12)
	25						***12 (9)***	***16 (9)***	24 (15)
	10						***14 (10)***	***16 (10)***	28 (17)
95 (89)	90						7 ***(3)***	8 (4)	11 (6)
	75						***7 (4)***	11 (6)	15 (9)
	50						***9 (7)***	14 ***(8)***	19 (12)
	25						***12 (9)***	***16 (9)***	23 (15)
	10						***14 (11)***	***16 (11)***	27 (18)
64 (57)	90						7 ***(3)***	8 (4)	10 (6)
	75						***7 (4)***	10 (6)	14 (8)
	50						***9 (7)***	14 ***(7)***	18 (11)
	25						***11 (9)***	17 ***(9)***	22 (14)
	10						***13 (10)***	19 ***(10)***	26 (17)

[a]Interpolate for intermediate frequencies. Values in the parentheses are for females. Numbers in bold italics, weight limited by the physiological design criterion (4 kcal/min for males; 3 kcal/min for females).

The maximum force that can be exerted by males with one hand (stronger hand) in the standing posture should not exceed 16 kg. For females, the maximum one-hand push exertion should not exceed 11 kg. If one-handed pushing exertions are to be applied repeatedly, these values should be reduced by 30% for both males and females up to 11 kg and 7.5 kg, respectively.

Chapter 6

Pulling

Pulling activity is similar to pushing activity in that the object weight is supported by the floor and force is applied to the object being pulled. As compared to manual lifting, pulling activities are not very strenuous. In this chapter, we provide data for designing two-handed and one-handed pulling activities.

Table 6.1 Recommended initial forces (kg) for male (female) industrial workers for two-handed pulling.

		Frequency of pull[a]							
Handle height (cm)	Population percentile	10/min	5/min	4/min	2.4/min	1/min	1/2 min	1/5 min	1/8 h
				2.1 m pulling distance					
144 (135)	90	14 (13)	16 (16)			18 (17)		19 (19)	23 (22)
	75	17 (16)	19 (19)			22 (20)		23 (23)	28 (26)
	50	20 (19)	23 (22)			26 (24)		28 (28)	33 (31)
	25	24 (21)	27 (25)			31 (28)		32 (32)	39 (35)
	10	26 (24)	30 (28)			34 (31)		36 (36)	44 (39)
95 (89)	90	19 (14)	22 (16)			25 (18)		27 (21)	32 (23)
	75	23 (16)	27 (19)			31 (21)		32 (25)	39 (27)
	50	28 (19)	32 (23)			36 (25)		39 (29)	47 (32)
	25	33 (22)	37 (26)			42 (29)		45 (33)	54 (37)
	10	37 (25)	42 (29)			48 (32)		51 (37)	61 (41)
64 (57)	90	22 (15)	25 (17)			28 (19)		30 (22)	36 (24)
	75	27 (17)	30 (20)			34 (22)		37 (26)	44 (28)
	50	32 (20)	36 (24)			41 (26)		44 (30)	53 (33)
	25	37 (23)	42 (27)			48 (30)		51 (35)	61 (38)
	10	42 (26)	48 (31)			54 (34)		57 (39)	69 (43)
				7.6 m pulling distance					
144 (135)	90			11 (11)		16 (16)		17 (17)	21 (20)
	75			14 (14)		20 (19)		21 (21)	26 (24)
	50			16 (16)		24 (22)		25 (25)	31 (28)
	25			19 (19)		28 (25)		29 (29)	36 (32)
	10			21 (21)		31 (28)		33 (32)	40 (36)
95 (89)	90			15 (14)		23 (16)		24 (19)	29 (21)
	75			19 (17)		28 (19)		29 (22)	36 (25)
	50			23 (19)		33 (23)		35 (26)	42 (29)
	25			26 (22)		39 (26)		41 (30)	49 (33)
	10			30 (25)		43 (29)		46 (33)	56 (37)
64 (57)	90			18 (15)		26 (17)		27 (20)	33 (22)
	75			21 (17)		31 (20)		33 (23)	40 (26)
	50			25 (20)		37 (24)		40 (28)	48 (30)
	25			30 (23)		44 (27)		46 (32)	56 (35)
	10			33 (26)		49 (31)		52 (35)	63 (39)
				15.2 m pulling distance					
144 (135)	90				13 (10)	15 (13)		16 (15)	20 (17)
	75				16 (12)	19 (16)		20 (18)	24 (20)
	50				19 (14)	22 (19)		24 (21)	29 (24)
	25				22 (16)	26 (21)		28 (25)	33 (27)
	10				24 (18)	29 (24)		31 (27)	38 (30)
95 (89)	90				18 (10)	21 (14)		23 (16)	28 (18)
	75				22 (12)	26 (17)		28 (19)	33 (21)
	50				26 (14)	31 (19)		33 (22)	40 (25)
	25				30 (16)	36 (22)		38 (26)	46 (28)
	10				33 (18)	41 (25)		43 (29)	52 (32)
64 (57)	90				20 (11)	24 (15)		26 (17)	31 (19)
	75				24 (13)	29 (17)		31 (20)	38 (22)
	50				29 (15)	35 (20)		37 (23)	45 (26)
	25				34 (17)	41 (23)		43 (27)	52 (30)
	10				38 (19)	46 (26)		49 (30)	59 (33)

Table 6.1 Recommended initial forces (kg) for male (female) industrial workers for two-handed pulling. (Continued)

		Frequency of pull[a]							
Handle height (cm)	Population percentile	10/min	5/min	4/min	2.4/min	1/min	1/2 min	1/5 min	1/8 h
					30.5 m pulling distance				
144 (135)	90					12 (12)		15 (14)	19 (17)
	75					14 (14)		19 (17)	23 (20)
	50					17 (17)		22 (20)	27 (24)
	25					20 (19)		26 (23)	32 (27)
	10					22 (22)		29 (25)	37 (31)
95 (89)	90					16 (13)		21 (15)	26 (18)
	75					20 (15)		26 (18)	32 (21)
	50					24 (18)		31 (21)	38 (25)
	25					27 (20)		36 (24)	45 (29)
	10					31 (23)		40 (26)	50 (32)
64 (57)	90					18 (13)		24 (15)	30 (19)
	75					22 (16)		29 (18)	36 (22)
	50					27 (18)		35 (22)	43 (26)
	25					31 (21)		41 (25)	50 (30)
	10					35 (24)		46 (28)	57 (34)
					45.7 m pulling distance				
144 (135)	90					10 (10)		13 (14)	16 (16)
	75					12 (12)		16 (17)	20 (20)
	50					15 (15)		19 (20)	24 (24)
	25					17 (17)		22 (23)	28 (27)
	10					20 (20)		25 (25)	31 (31)
95 (89)	90					14 (13)		18 (15)	23 (18)
	75					17 (15)		22 (18)	28 (21)
	50					20 (18)		27 (21)	33 (25)
	25					24 (20)		31 (24)	38 (29)
	10					27 (23)		35 (26)	43 (32)
64 (57)	90					16 (13)		21 (15)	26 (19)
	75					19 (16)		25 (18)	31 (22)
	50					23 (18)		30 (22)	37 (26)
	25					27 (21)		35 (25)	43 (30)
	10					30 (24)		39 (28)	49 (34)
					61 m pulling distance				
144 (135)	90						10 (10)	11 (11)	14 (14)
	75						12 (12)	14 (14)	17 (17)
	50						14 (14)	16 (16)	20 (20)
	25						16 (16)	19 (19)	24 (24)
	10						18 (18)	21 (21)	27 (27)
95 (89)	90						13 (12)	16 (13)	19 (16)
	75						16 (15)	19 (16)	24 (19)
	50						20 (17)	23 (18)	28 (22)
	25						23 (20)	26 (21)	33 (26)
	10						26 (22)	30 (24)	37 (29)
64 (57)	90						15 (13)	18 (14)	22 (17)
	75						19 (15)	21 (16)	27 (20)
	50						22 (18)	26 (19)	32 (23)
	25						26 (21)	30 (22)	37 (27)
	10						29 (23)	34 (25)	42 (30)

[a]Interpolate for intermediate frequencies. Values in parentheses are for females.

Table 6.2 Recommended sustained forces (kg) for male (female) industrial workers for two-handed pulling

		Frequency of pull[a]							
Handle height (cm)	Population percentile	10/min	5/min	4/min	2.4/min	1/min	1/2 min	1/5 min	1/8 h
		2.1 m pulling distance							
144 (135)	90	8 ***(5)***	10 ***(8)***			12 (10)		15 (11)	18 (15)
	75	10 ***(7)***	13 ***(10)***			16 (13)		19 (15)	23 (20)
	50	***10 (9)***	16 ***(14)***			20 (17)		23 (19)	28 (25)
	25	***12 (11)***	20 ***(16)***			24 (21)		28 (23)	34 (31)
	10	***14 (13)***	22 ***(19)***			27 (24)		32 (27)	39 (36)
95 (89)	90	10 ***(5)***	13 ***(8)***			16 (10)		19 (11)	24 (14)
	75	***10 (7)***	17 ***(10)***			21 (13)		25 (15)	30 (19)
	50	***13 (9)***	21 ***(13)***			26 (16)		31 (19)	37 (25)
	25	***15 (10)***	***21 (15)***			31 (20)		37 (23)	45 (30)
	10	***18 (12)***	***23 (18)***			36 (23)		42 (26)	51 (35)
64 (57)	90	11 ***(4)***	14 (8)			17 (9)		20 (10)	25 (13)
	75	***11 (6)***	19 ***(9)***			23 (12)		26 (13)	32 (18)
	50	***14 (8)***	***18 (12)***			28 (15)		32 (17)	40 (23)
	25	***16 (9)***	***21 (15)***			33 (18)		39 (21)	48 (27)
	10	***18 (11)***	***25 (17)***			38 (21)		45 (24)	54 (32)
		7.6 m pulling distance							
144 (135)	90			6 ***(6)***		10 (9)		12 (10)	15 (13)
	75			***6 (6)***		13 (12)		16 (13)	19 (18)
	50			***8 (8)***		16 ***(13)***		19 (17)	23 (22)
	25			***10 (10)***		20 ***(15)***		23 (21)	28 (27)
	10			***11 (11)***		22 ***(18)***		26 (24)	32 (32)
95 (89)	90			***6 (6)***		13 (9)		16 (10)	19 (13)
	75			***9 (7)***		17 (11)		20 (13)	25 (17)
	50			***10 (9)***		21 ***(13)***		25 (16)	31 (22)
	25			***13 (12)***		26 ***(15)***		30 (20)	37 (27)
	10			***14 (14)***		29 ***(18)***		34 (23)	42 (31)
64 (57)	90			***7 (5)***		14 (8)		17 (9)	20 (12)
	75			***9 (7)***		19 (11)		22 (12)	26 (16)
	50			***11 (9)***		23 (13)		27 (15)	33 (20)
	25			***14 (11)***		27 ***(14)***		32 (19)	39 (24)
	10			***15 (13)***		***25 (16)***		37 (22)	45 (28)
		15.2 m pulling distance							
144 (135)	90				***6 (4)***	9 ***(6)***		10 (8)	13 (11)
	75				***7 (6)***	12 ***(9)***		14 (11)	17 (15)
	50				***9 (6)***	14 ***(11)***		17 (14)	20 (19)
	25				***10 (9)***	***14 (13)***		20 (17)	24 (23)
	10				***11 (11)***	***15 (15)***		23 (20)	28 (27)
95 (89)	90				***7 (4)***	12 ***(6)***		14 (8)	17 (11)
	75				***9 (6)***	15 ***(9)***		18 (11)	22 (14)
	50				***11 (8)***	***15 (10)***		22 (14)	27 (18)
	25				***14 (9)***	***18 (13)***		26 (17)	32 (22)
	10				***15 (11)***	***21 (15)***		30 (20)	37 (26)
64 (57)	90				***7 (4)***	12 ***(6)***		15 (7)	18 (10)
	75				***10 (6)***	***13 (8)***		19 (10)	23 (13)
	50				***12 (7)***	***16 (9)***		23 (13)	28 (17)
	25				***14 (9)***	***19 (12)***		28 (16)	34 (21)
	10				***16 (10)***	***22 (14)***		32 (18)	39 (24)

Table 6.2 Recommended sustained forces (kg) for male (female) industrial workers for two-handed pulling (Continued)

Handle height (cm)	Population percentile	Frequency of pull[a] 10/min	5/min	4/min	2.4/min	1/min	1/2 min	1/5 min	1/8 h
				30.5 m pulling distance					
144 (135)	90					7 ***(5)***		9 (7)	13 (10)
	75					***7 (7)***		12 (10)	16 (14)
	50					***9 (9)***		15 (12)	20 (17)
	25					***10 (10)***		18 ***(13)***	24 (21)
	10					***12 (12)***		20 ***(15)***	27 (25)
95 (89)	90					***7 (5)***		12 (7)	17 (10)
	75					***10 (7)***		16 (9)	21 (13)
	50					***11 (9)***		19 (12)	26 (17)
	25					***14 (10)***		23 ***(13)***	32 (21)
	10					***15 (12)***		27 ***(15)***	36 (24)
64 (57)	90					***7 (5)***		13 (6)	18 (9)
	75					***10 (6)***		17 (9)	23 (12)
	50					***12 (8)***		21 (11)	27 (16)
	25					***14 (9)***		25 (13)	33 (19)
	10					***17 (11)***		28 ***(14)***	38 (22)
				45.7 m pulling distance					
144 (135)	90					***5 (5)***		8 (7)	10 (9)
	75					***6 (6)***		10 (9)	14 (12)
	50					***7 (7)***		12 ***(9)***	17 (16)
	25					***9 (9)***		15 ***(12)***	20 (19)
	10					***10 (10)***		17 ***(14)***	23 (23)
95 (89)	90					***6 (4)***		10 (6)	14 (9)
	75					***8 (6)***		13 (9)	18 (12)
	50					***10 (8)***		16 ***(9)***	22 (15)
	25					***11 (9)***		19 ***(11)***	26 (19)
	10					***13 (11)***		22 ***(14)***	30 (22)
64 (57)	90					***6 (4)***		11 (6)	15 (8)
	75					***8 (6)***		14 (8)	19 (11)
	50					***10 (8)***		17 ***(9)***	23 (14)
	25					***12 (9)***		21 ***(10)***	28 (17)
	10					***14 (10)***		24 ***(12)***	32 (20)
				61 m pulling distance					
144 (135)	90						6 ***(4)***	6 (5)	9 (7)
	75						7 ***(6)***	8 (7)	11 (10)
	50						***7 (7)***	10 ***(8)***	14 (12)
	25						***9 (8)***	12 ***(9)***	17 (15)
	10						***10 (10)***	14 ***(10)***	19 (17)
95 (89)	90						7 ***(4)***	9 (5)	12 (7)
	75						***7 (5)***	11 (7)	15 (9)
	50						***10 (7)***	14 ***(7)***	18 (12)
	25						***11 (8)***	16 ***(8)***	22 (15)
	10						***13 (10)***	19 ***(10)***	25 (17)
64 (57)	90						8 ***(3)***	9 (5)	12 (6)
	75						***8 (5)***	12 (6)	16 (9)
	50						***10 (7)***	14 ***(7)***	20 (11)
	25						***12 (8)***	17 ***(8)***	23 (13)
	10						***14 (9)***	***16 (9)***	27 (16)

[a]Interpolate for intermediate frequencies. Values in the parentheses are for females. Numbers in bold italics, weight limited by the physiological design criterion (4 kcal/min for males; 3 kcal/min for females).

Two-handed pulling

For a frequent pulling task, the biomechanical design criterion is not limiting. Therefore, we modified the psychophysical design data so that the physiological design criterion was not violated. Tables 6.1 and 6.2 provide initial and sustained pulling forces for two-handed pulling tasks for different pulling distances. As shown in Figure 6.1, pulling force is maximized when the handle height is approximately 30–40% of the reach height. A foot distance of −10% reach height is optimum; a negative value of the foot distance means the foot is located on the other side of the vertical plane of the handle.

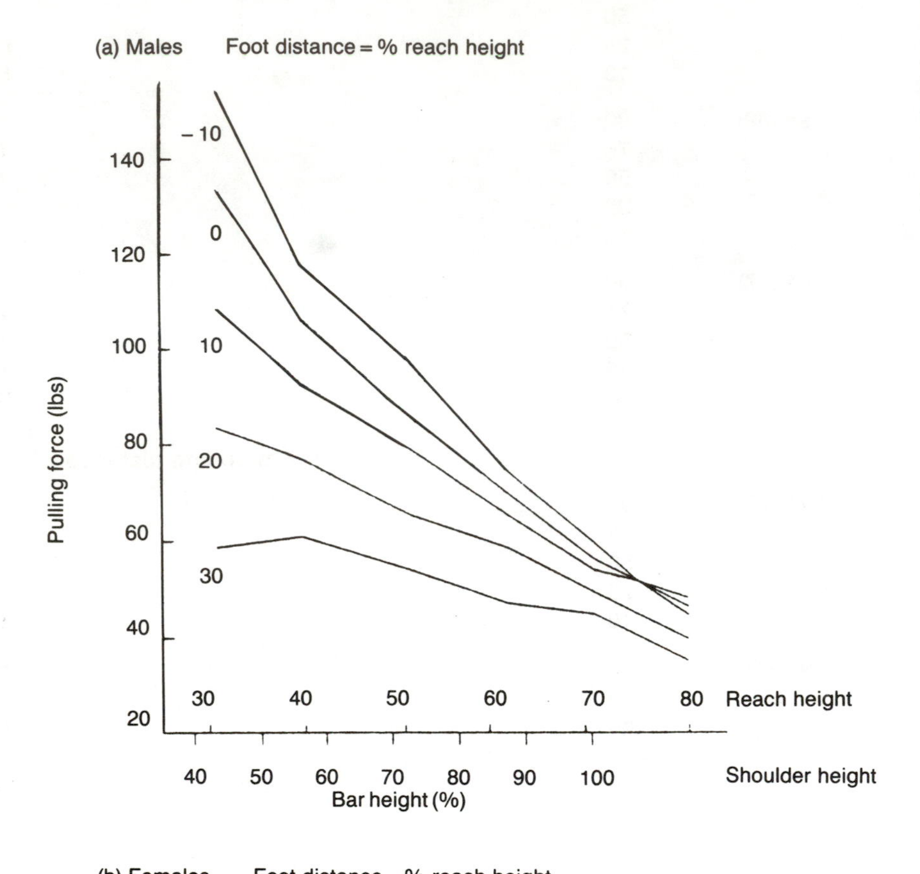

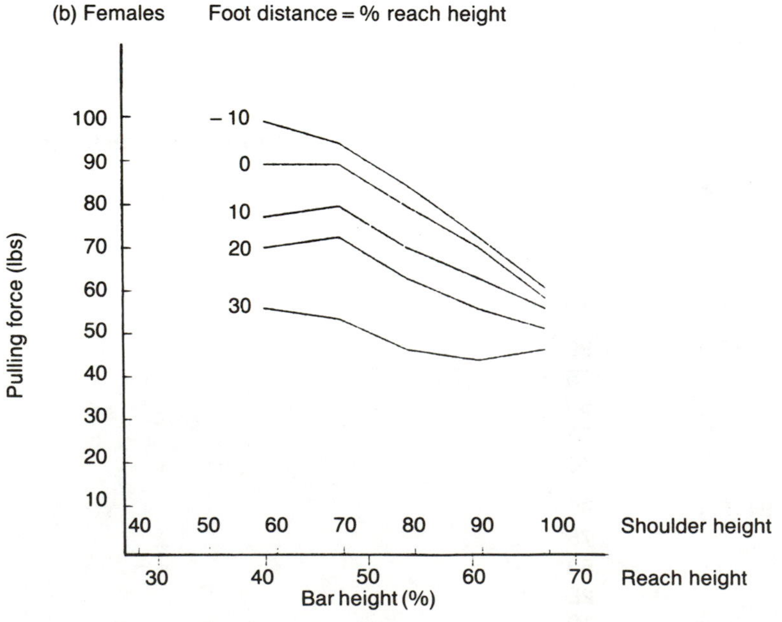

Figure 6.1 Isometric pulling force (kg) for different foot distances and handle heights.

One-handed pulling

The maximum horizontal pull force that can be exerted by males with one hand (stronger hand), infrequently, is approximately 15 kg on average. Females, on average, are expected to exert approximately 10 kg force with their stronger hand. For frequent exertions, the maximum one-handed pull force should not exceed 10 kg for males and 7 kg for females.

Chapter 7

Carrying

Carrying activities, even though less pervasive than manual lifting, are still being performed routinely in industry. The carrying distances, however, are usually short. The carrying capacity of workers is generally influenced by the same factors that influence their lifting capacity. There are, however, some exceptions. For instance, load

Table 7.1 Recommended weight of carry (kg) for male industrial workers for two-handed carrying of symmetrical loads for 8 h.

		Frequency of carrying[a]						
Carrying height (cm)	Population percentile	10/min	6/min	5/min	3.3/min	1/min	1/5 min	1/8 h
				2.1 m carrying distance				
111	90	10		14		17	19	25
	75	14		19		23	26	27[b]
	50	***15***		25		27[b]	27[b]	27[b]
	25	***18***		27[b]		27[b]	27[b]	27[b]
	10	***21***		27[b]		27[b]	27[b]	27[b]
79	90	13		17		21	23	27[b]
	75	18		23		27[b]	27[b]	27[b]
	50	***18***		27[b]		27[b]	27[b]	27[b]
	25	***22***		27[b]		27[b]	27[b]	27[b]
	10	***26***		27[b]		27[b]	27[b]	27[b]
				4.3 m carrying distance				
111	90		9			15	17	22
	75		13			21	23	27[b]
	50		16			27	27[b]	27[b]
	25		***16***			27[b]	27[b]	27[b]
	10		***19***			27[b]	27[b]	27[b]
79	90		11			18	21	27
	75		16			25	27[b]	27[b]
	50		***16***			27[b]	27[b]	27[b]
	25		***20***			27[b]	27[b]	27[b]
	10		***23***			27[b]	27[b]	27[b]
				8.5 m carrying distance				
111	90				10	13	15	20
	75				13	18	20	27
	50				17	23	26	27[b]
	25				***17***	27[b]	27[b]	27[b]
	10				***19***	27[b]	27[b]	27[b]
79	90				13	17	20	26
	75				17	24	27	27[b]
	50				***17***	27[b]	27[b]	27[b]
	25				***22***	27[b]	27[b]	27[b]
	10				***25***	27[b]	27[b]	27[b]

[a]Interpolate for intermediate frequencies. [b]Weight limited by the biomechanical design criterion (3930 N spinal compression). Number in bold italics, weight limited by the physiological design criterion (4 kcal/min).

dimension in the sagittal plane (dimension away from the body) does not influence carrying capacity to the same extent as it does the lifting capacity. The decline in carrying capacity, as the load dimension in the sagittal plane increases from 30 to 61 cm, has been observed to be less than 4%. This has led to the conclusion that load dimension in the sagittal plane is not a very important factor in determining manual carrying capabilities of workers; **it is not the load size but the operator's anthropometry that is the limiting factor.** The influence of many other task factors, such as load asymmetry and couplings, on carrying capabilities is very similar to their effect on lifting capabilities.

Table 7.2 Recommended weight of carry (kg) for female industrial workers for two-handed carrying of symmetrical loads for 8 h.

		Frequency of carrying[a]						
Carrying height (cm)	Population percentile	10/min	6/min	5/min	3.3/min	1/min	1/5 min	1/8 h
				2.1 m carrying distance				
105	90	11		12		13	13	18
	75	***11***		14		15	16	20[b]
	50	***13***		15		18	18	20[b]
	25	***15***		***15***		20	20[b]	20[b]
	10	***16***		***17***		20[b]	20[b]	20[b]
72	90	***11***		14		16	16	20[b]
	75	***13***		***15***		18	19	20[b]
	50	***15***		***16***		20[b]	20[b]	20[b]
	25	***17***		***19***		20[b]	20[b]	20[b]
	10	***19***		20[b]		20[b]	20[b]	20[b]
				4.3 m carrying distance				
105	90		9			13	13	18
	75		***9***			15	16	20[b]
	50		***10***			18	18	20[b]
	25		***12***			20	20[b]	20[b]
	10		***14***			20[b]	20[b]	20[b]
72	90		10			14	14	20
	75		11			16	17	20[b]
	50		***11***			19	20	20[b]
	25		***13***			20[b]	20[b]	20[b]
	10		***15***			20[b]	20[b]	20[b]
				8.5 m carrying distance				
105	90				10	12	12	16
	75				***10***	14	14	19
	50				***12***	16	16	20[b]
	25				***13***	18	19	20[b]
	10				***15***	20	20[b]	20[b]
72	90				***10***	14	14	19
	75				***12***	16	17	20[b]
	50				***14***	19	20	20[b]
	25				***15***	20[b]	20[b]	20[b]
	10				***17***	20[b]	20[b]	20[b]

[a]Interpolate for intermediate frequencies. [b]Weight limited by the biomechanical design criterion (2689 N spinal compression). Number in bold italics, weight limited by the physiological design criterion (3 kcal/min).

Two-handed carrying

Since the load is entirely supported by the operator, determination of safe carrying load must be based on the same design criteria as the determination of safe lifting load. Just as in the case of manual lifting, we began with the psychophysical data and adjusted it where the psychophysical weight recommendation violated either the biomechanical or the physiological design criterion. **Tables 7.1 and 7.2 provide the basic carrying capability design database for males and females, respectively. These data must be adjusted to reflect the influence of realistic task conditions. Adjustment multipliers for working duration, limited headroom, asymmetrical handling (turning while lifting to carry the load), load asymmetry, couplings and heat stress should be taken from Tables 4.4, 4.5, 4.6, 4.7, 4.8 and 4.10, respectively. The design data in Tables 7.1 and 7.2 should be considered for a load approximately 30 cm wide in the sagittal plane. For a 61-cm wide load, the carrying capability should be reduced by 4%; linear interpolation may be used for intermediate load dimensions in the sagittal plane. The final carrying capability is determined in the same way as the lifting capacity – by multiplying recommended weight from Tables 7.1 and 7.2 by the multipliers from Tables 4.4 to 4.8 and Table 4.10. It is also recommended that weight be carried at the lower height (79 cm for men and 72 cm for women – approximately the knuckle height).**

One-handed carrying

Table 7.3 provides the recommended weight that should be carried in one hand (stronger hand) infrequently. **If carrying is to be performed frequently, the recommended weight should be reduced by 30%.**

Table 7.3 Recommended weight of carry (kg) for one-handed infrequent carrying

Carrying distance (m)[a]	Population percentile	Males	Females
30.48	90	6.5	5.5
	75	8.5	7
	50	11	8
	25	13.5	9
	10	15.5	10.5
60.96	90	6	5.5
	75	8	6.5
	50	10	7.5
	25	12	8.5
	10	14	9.5
91.44	90	6	5
	75	7.5	6
	50	9	7
	25	10.5	8
	10	12	9

[a]Interpolate for intermediate distances.

Chapter 8

Holding

While holding is undesirable as it leads to static fatigue, occasionally there is a need to hold an object in place. This chapter provides holding time data for loads of different weights for both males and females (data from Ayoub *et al.*, 1987). In all, data for 55 variations of four basic holding activities are included. These four basic activities are: (1) holding a container against a vertical surface (wall), no barrier, activity code WNU; (2) holding a container against wall with a vertical restraining barrier (1.9 × 152 mm cross-section board located at a horizontal distance of 38 cm from the target, towards the subject), activity code WNC; (3) holding a container against wall with an overhead restraining barrier (ceiling), activity code WCU; and (4) holding a container against ceiling, activity code OVU. For each activity, a variety of body postures are included. The basic postures are: standing (code ST), sitting (0.3 m high seat, code SI), squatting (code SQ), kneeling (code KN), and lying (code LY).

The holding task is described by a seven-digit alphanumeric character and consisted of holding the container against a target (located at a specified percentage of the standing vertical reach (SVR) for holding against a vertical surface, located at a specified percentage of the SVR plus 0.25 m (height of the container) for holding against the ceiling). The first three characters describe the activity (one of the four described above); the next two characters describe the posture (one of the five described above); the sixth character describes the height of the target (object) as a percentage of the SVR; the seventh character describes the height of the ceiling as a percentage of the SVR. A '.' in the task code indicates that no vertical target is present under the OVU holding conditions.

For those tasks in which the container is held against a vertical surface (WNU, WNC, and WCU tasks), the vertical target is such that the lower edge of the target is at the specified SVR.

Table 8.1 provides codes for each of the 55 holding tasks and the associated description. Figure 8.1 shows a holding task where the container is held against the wall at 20% SVR height, subject is in the kneeling posture, and there is no ceiling and no constraint. Figure 8.2 shows a standing holding task, container held against the wall with no ceiling and vertical constraint. Figure 8.3 shows sitting holding task, container held against the wall and overhead ceiling constraint. Figure 8.4 shows a standing holding task, container held against the ceiling. Table 8.2 shows the holding times for different loads for males and females.

Table 8.1 Description of codes and holding tasks

Task	Description
WNUST8.	Container held against wall, no ceiling, no constraint, standing and target at 80% of SVR
WNUST6.	Container held against wall, no ceiling, no constraint, standing and target at 60% of SVR
WNUST4.	Container held against wall, no ceiling, no constraint, standing and target at 40% of SVR
WNUSI6.	Container held against wall, no ceiling, no constraint, sitting and target at 60% of SVR
WNUSI4.	Container held against wall, no ceiling, no constraint, sitting and target at 40% of SVR
WNUSI2.	Container held against wall, no ceiling, no constraint, sitting and target at 20% of SVR
WNUSQ6.	Container held against wall, no ceiling, no constraint, squatting and target at 60% of SVR
WNUSQ4.	Container held against wall, no ceiling, no constraint, squatting and target at 40% of SVR
WNUSQ2.	Container held against wall, no ceiling, no constraint, squatting and target at 20% of SVR
WNUKN6.	Container held against wall, no ceiling, no constraint, kneeling and target at 60% of SVR
WNUKN4.	Container held against wall, no ceiling, no constraint, kneeling and target at 40% of SVR
WNUKN2.	Container held against wall, no ceiling, no constraint, kneeling and target at 20% of SVR
WNULY3.	Container held against wall, no ceiling, no constraint, lying and target at 30% of SVR
WNULY2.	Container held against wall, no ceiling, no constraint, lying and target at 20% of SVR
WNCST8.	Container held against wall, no ceiling, constraint, standing and target at 80% of SVR
WNCST6.	Container held against wall, no ceiling, constraint, standing and target at 60% of SVR
WNCST4.	Container held against wall, no ceiling, constraint, standing and target at 40% of SVR
WNCSI6.	Container held against wall, no ceiling, constraint, sitting and target at 60% of SVR
WNCSI4.	Container held against wall, no ceiling, constraint, sitting and target at 40% of SVR
WNCSI2.	Container held against wall, no ceiling, constraint, sitting and target at 20% of SVR
WNCSQ6.	Container held against wall, no ceiling, constraint, squatting and target at 60% of SVR
WNCSQ4.	Container held against wall, no ceiling, constraint, squatting and target at 40% of SVR
WNCSQ2.	Container held against wall, no ceiling, constraint, squatting and target at 20% of SVR
WNCKN6.	Container held against wall, no ceiling, constraint, kneeling and target at 60% of SVR
WNCKN4.	Container held against wall, no ceiling, constraint, kneeling and target at 40% of SVR
WNCKN2.	Container held against wall, no ceiling, constraint, kneeling and target at 20% of SVR
WCUST56	Container held against wall, ceiling constraint, standing, target at 60% of SVR and ceiling at 50% of SVR
WCUST46	Container held against wall, ceiling constraint, standing, target at 60% of SVR and ceiling at 40% of SVR
WCUSI34	Container held against wall, ceiling constraint, sitting, target at 40% of SVR and ceiling at 30% of SVR
WCUSI24	Container held against wall, ceiling constraint, sitting, target at 40% of SVR and ceiling at 20% of SVR
WCUSI23	Container held against wall, ceiling constraint, sitting, target at 30% of SVR and ceiling at 20% of SVR
WCUSQ34	Container held against wall, ceiling constraint, squatting, target at 40% of SVR and ceiling at 30% of SVR
WCUSQ24	Container held against wall, ceiling constraint, squatting, target at 40% of SVR and ceiling at 20% of SVR
WCUSQ23	Container held against wall, ceiling constraint, squatting, target at 30% of SVR and ceiling at 20% of SVR
WCUKN34	Container held against wall, ceiling constraint, kneeling, target at 40% of SVR and ceiling at 30% of SVR
WCUKN24	Container held against wall, ceiling constraint, kneeling, target at 40% of SVR and ceiling at 20% of SVR
WCUKN23	Container held against wall, ceiling constraint, kneeling, target at 30% of SVR and ceiling at 20% of SVR
OVUST.8	Container held against ceiling, no constraint, standing and target at 80% of SVR
OVUST.7	Container held against ceiling, no constraint, standing and target at 70% of SVR
OVUST.6	Container held against ceiling, no constraint, standing and target at 60% of SVR
OVUSI.6	Container held against ceiling, no constraint, sitting and target at 60% of SVR
OVUSI.5	Container held against ceiling, no constraint, sitting and target at 50% of SVR
OVUSI.4	Container held against ceiling, no constraint, sitting and target at 40% of SVR
OVUSI.3	Container held against ceiling, no constraint, sitting and target at 30% of SVR
OVUSQ.6	Container held against ceiling, no constraint, squatting and target at 60% of SVR
OVUSQ.5	Container held against ceiling, no constraint, squatting and target at 50% of SVR
OVUSQ.4	Container held against ceiling, no constraint, squatting and target at 40% of SVR
OVUSQ.3	Container held against ceiling, no constraint, squatting and target at 30% of SVR
OVUKN.7	Container held against ceiling, no constraint, kneeling and target at 70% of SVR
OVUKN.6	Container held against ceiling, no constraint, kneeling and target at 60% of SVR
OVUKN.5	Container held against ceiling, no constraint, kneeling and target at 50% of SVR
OVUKN.4	Container held against ceiling, no constraint, kneeling and target at 40% of SVR
OVUKN.3	Container held against ceiling, no constraint, kneeling and target at 30% of SVR
OVULY.3	Container held against ceiling, no constraint, lying and target at 30% of SVR
OVULY.2	Container held against ceiling, no constraint, lying and target at 20% of SVR

Table 8.2 Average holding time[a] (s) for different loads (kg) for males and females

	Male					Female				
Task	9 kg	18 kg	27 kg	36 kg	45 kg	9 kg	18 kg	27 kg	36 kg	45 kg
WNUST8	60	55	35	15	8	53.5	17	1	1	1
WNUST6	60	59	55	49	20	60	51.5	20.5	4	1
WNUST4	60	60	60	42	22	60	44	19	8	1
WNUSI6	60	47	23	7	3	47	4	0	0	0
WNUSI4	60	57	51	38	17	58	38.5	13	0	0
WNUSI2	60	60	55	51	42	60	50	25	10.5	1
WNUSQ6	58.5	53.5	31.5	10.5	3	47	10.5	0	0	0
WNUSQ4	58	55	44	23.5	12	50	25.5	5	0	0
WNUSQ2	60	60	57	40.5	23	56.5	38.5	12	3.5	0
WNUKN6	55	47	16.5	8.5	3.5	37	3.5	0	0	0
WNUKN4	60	57	50	35	29	56.5	42	7	0	0
WNUKN2	60	60	58	53	38	55	48.5	28.5	8.5	4
WNULY3	58.5	35	15	6.5	1	37	7	0	0	0
WNULY2	55	36.5	18	3.5	0	45	12	0	0	0
WNCST8	60	51.5	28.5	7	3	51.5	8	0	0	0
WNCST6	60	58.5	52	36.5	17	58.5	38.5	8	0	0
WNCST4	57	37	15	3	3	43.5	8.5	0	0	0
WNCSI6	60	48.5	16.5	3.5	3.5	58.5	9	0	0	0
WNCSI4	56.5	40	15	3.5	1.5	46.5	8.5	0	0	0
WNCSI2	58	38.5	15	8.5	3	48.5	8	0	0	0
WNCSQ6	58.5	47	15	5	3	46.5	3	0	0	0
WNCSQ4	55	30	10	1.5	0	26	3	0	0	0
WNCSQ2	60	36.5	13	5	0	42	8	0	0	0
WNCKN6	55	43.5	22	7	3	37	3	0	0	0
WNCKN4	60	53	43.5	23	7	53.5	31.5	15	0	0
WNCKN2	58.5	30	15	3	0	38.5	3	0	0	0
WCUST56	60	60	48	30	10	56.5	16.5	0	0	0
WCUST46	60	57	38	30	18.5	53	25	7	3.5	0
WCUSI34	60	60	43	33	18	53	32	3.5	0	0
WCUSI24	60	55	47	40	35.5	55	48.5	43.5	9	3
WCUSI23	60	50	33	12	3	55	18.5	0	0	0
WCUSQ34	58	53	43.5	32	15	58	33.5	7	0	0
WCUSQ24	60	58	55	43	35	53	38	11.5	3	0
WCUSQ23	58.5	50	33	13	3	53.5	20	7	0	0
WCUKN34	60	53	33	21	8	53	28	3	0	0
WCUKN24	60	55	40	35	25	56.5	36.5	18.5	3	0
WCUKN23	58	56.5	33	20	7	55	22	3	0	0
OVUST.8	50	31	15	7	3	16.5	3	0	0	0
OVUST.7	51.5	23	7	3	1	18	3	0	0	0
OVUST.6	60	55	41	21	13.5	58	28	2	0	0
OVUSI.6	48.5	20	10	5	4	11.5	0	0	0	0
OVUSI.5	41.5	10	1.5	0	0	8.5	0	0	0	0
OVUSI.4	58	53	33	12	4	47	15	2	0	0
OVUSI.3	53	36.5	23	20	6.5	21	3	0	0	0
OVUSQ.6	46.5	20	8.5	3.5	2	21	5	0	0	0
OVUSQ.5	41.5	8	2	0	0	5	0	0	0	0
OVUSQ.4	55	25	9	3	3	16.5	2	0	0	0
OVUSQ.3	57	45	26.5	10	3	43.5	10	0	0	0
OVUKN.7	55	33	13.5	6.5	3	26.5	7	0	0	0
OVUKN.6	42	15	5	1.5	0	8	0	0	0	0
OVUKN.5	48.5	15	3.5	0	0	11.5	0	0	0	0
OVUKN.4	58	46.5	23	8	1	45	15	0	0	0
OVUKN.3	55	36.5	23	11.5	7	35	13	3.5	0	0
OVULY.3	60	38.5	28	10	3	48	46.5	0	0	0
OVULY.2	48.5	15	1.5	0	0	10	0	0	0	0

[a]Holding time not recorded beyond 60 s.

Figure 8.1 Holding task WNUSQ2 – container held against wall, no ceiling, no constraint, squatting and target at 20% of SVR.

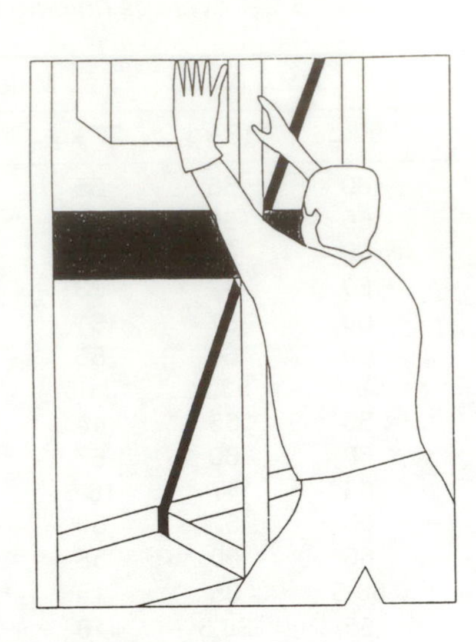

Figure 8.2 Holding task WNCST8 – container held against wall, no ceiling, constraint, standing and target at 80% of SVR.

Figure 8.3 Holding task WCUSI24 – container held against wall, ceiling constraint, sitting, target at 40% of SVR and ceiling at 20% of SVR.

Figure 8.4 Holding task OVUST.6 – container held against ceiling, no constraint, standing and target at 60% of SVR.

Reference

Ayoub, M.M., Smith, J.L., Selan, J.L., Chen, H.C., Lee, Y.H., Kim, H.K. and Fernandez, J.E., 1987. Manual material handling in unusual posture. Technical Report, Department of Industrial Engineering, Texas Tech University, Lubbock, Texas, USA.

Chapter 9

Material handling in unusual postures

Frequently, MMH tasks must be performed in unusual postures. These tasks are performed occasionally, but not frequently. Very little data are available to help design MMH tasks that are performed in unusual postures. Furthermore, the available data are based on either the psychophysical design approach or the static strength approach. Here we present design data for lifting, pushing and pulling in unusual postures. The lifting data conform to the psychophysical and biomechanical design criteria. The push and pull data are based on maximum volitional isometric (static) strength.

Lifting

Figures 9.1–9.19 show two-handed lifting in unusual postures. Figures 9.20–9.28 show one-handed lifting in unusual postures (data from Ayoub *et al.*, 1987). All two-handed sagittal lifting tasks involve lifting to three different height levels: 35%, 60%, and 85% of vertical reach height. All one-handed sagittal lifting tasks also involve lifting to three different height levels: 35%, 60%, and 85% of vertical reach minus 25 cm (distance between the handles and the container bottom). The design data for males and females are given in Tables 9.1 and 9.2, respectively.

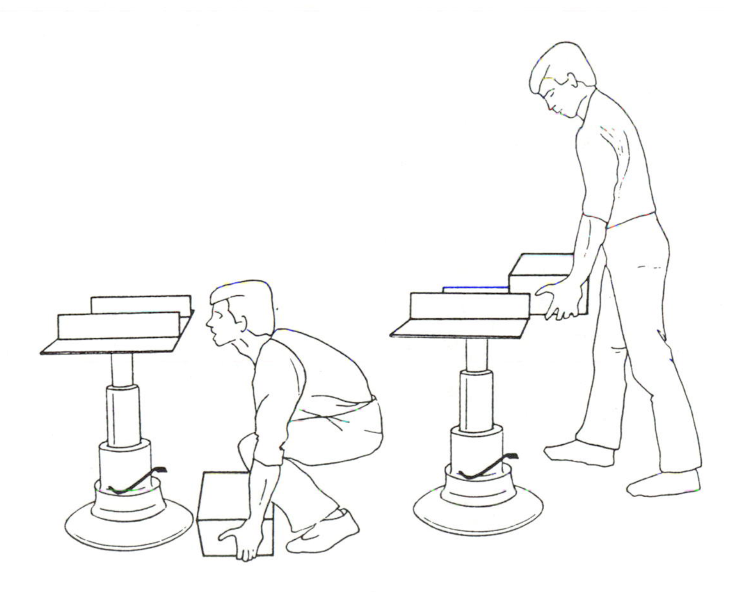

Figure 9.1 Standing lift with a 61 × 30.5 × 15 cm box to 35, 60 and 85% of vertical reach heights (knees bent, back straight, divider untouched) – tasks A1, A2, and A3, respectively.

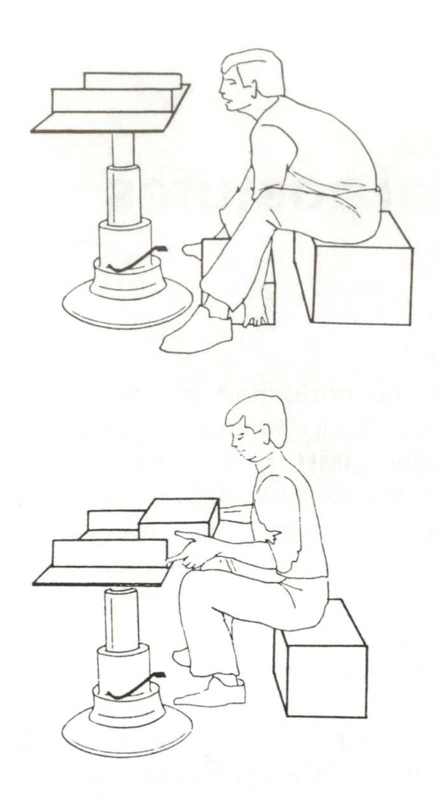

Figure 9.2 Sitting lift with a 61 × 30.5 × 15 cm box, sitting on a 30.5 cm high chair, to 35, 60, and 80% vertical reach heights (divider untouched) – tasks B1, B2, and B3, respectively.

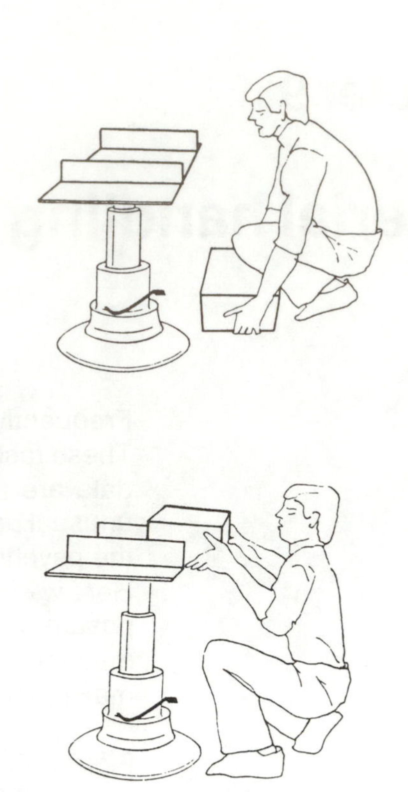

Figure 9.3 Squatting lift with a 61 × 30.5 × 15 cm box to 35, 60, and 80% vertical reach heights (dividers untouched) – tasks C1, C2, and C3, respectively.

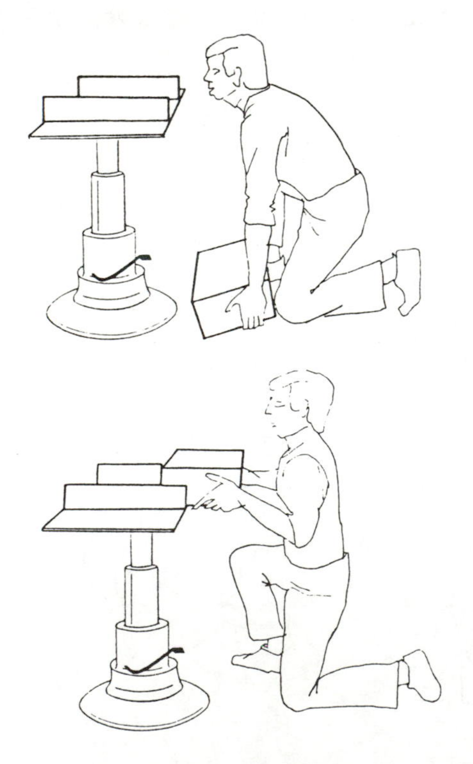

Figure 9.4 One knee kneeling lift with a 61 × 30.5 × 15 cm box to 35, 60, and 80% vertical reach heights (left knee down, right knee up, divider untouched) – tasks D1, D2, and D3, respectively.

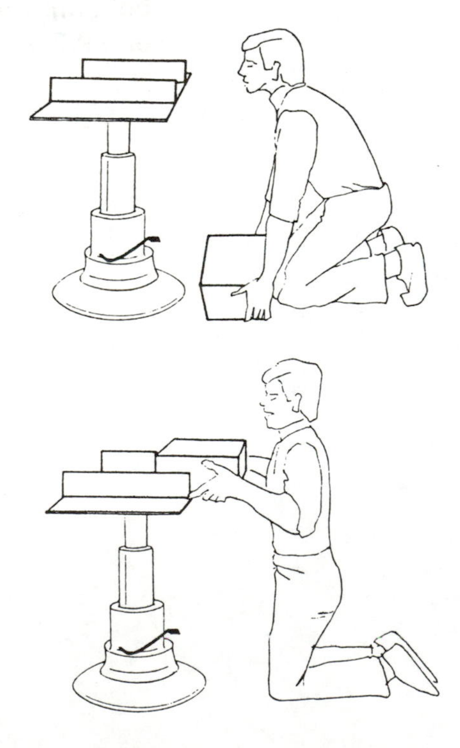

Figure 9.5 Two knees kneeling lift with a 61 × 30.5 × 15 cm box to 35, 60, and 80% vertical reach heights (divider untouched) – tasks E1, E2, and E3, respectively.

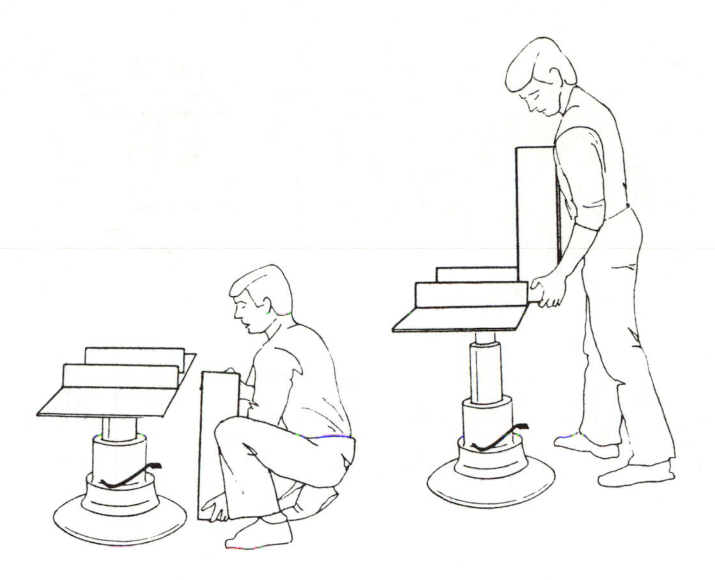

Figure 9.6 Standing lift with a 30.5 × 15 × 61 cm box to 35, 60, and 80% vertical reach heights (knees bent, back straight, divider untouched) – tasks F1, F2, and F3, respectively.

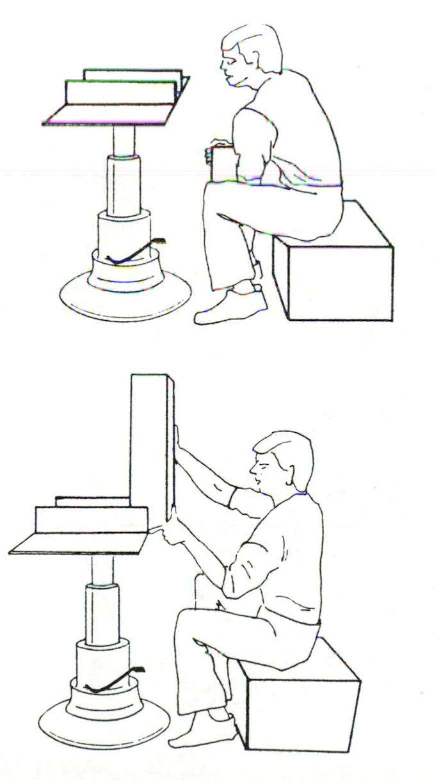

Figure 9.7 Sitting lift with a 30.5 × 15 × 61 cm box, sitting on a 30.5 cm high chair, to 35, 60, and 85% vertical reach heights (divider untouched) – tasks G1, G2, and G3, respectively.

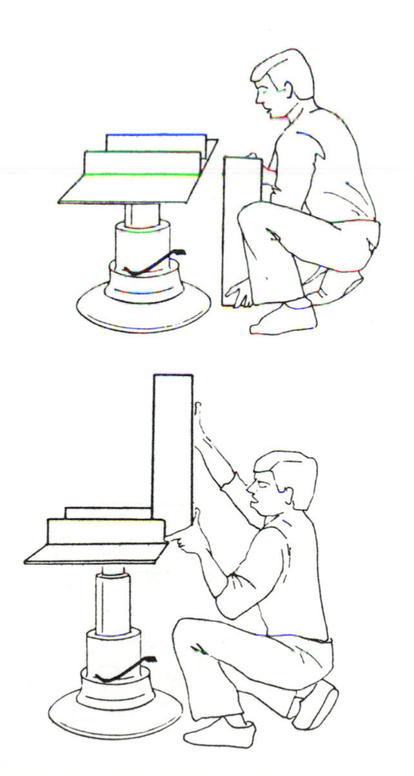

Figure 9.8 Squatting lift with a 30.5 × 15 × 61 cm box to 35, 60, and 85% vertical reach heights (divider untouched) – tasks H1, H2, and H3, respectively.

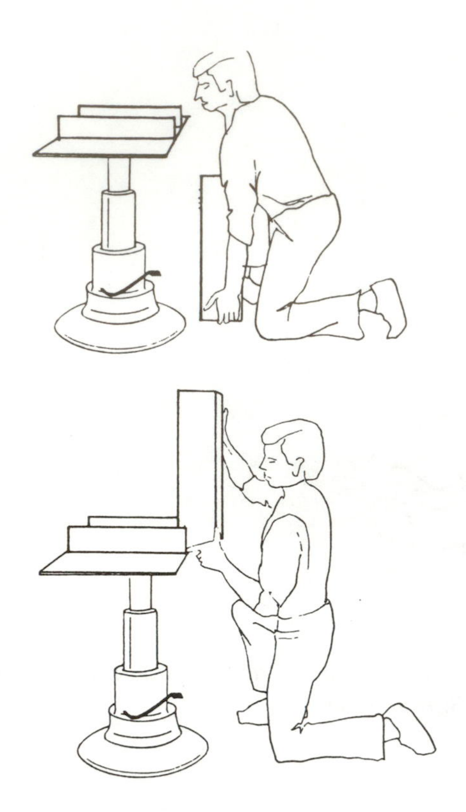

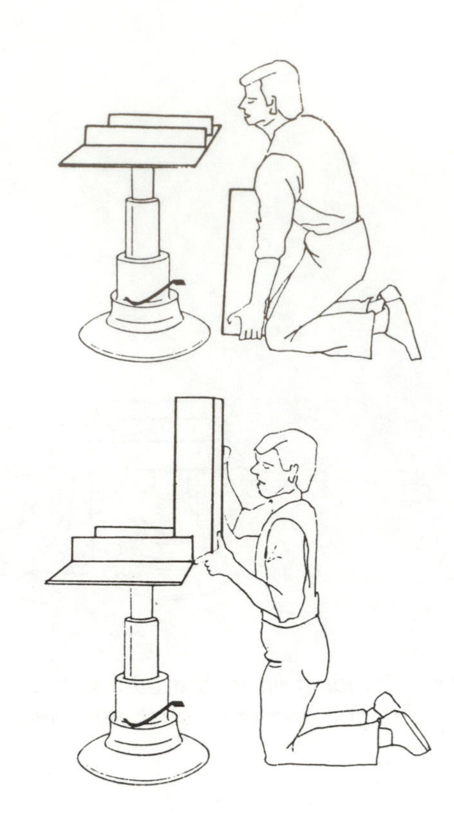

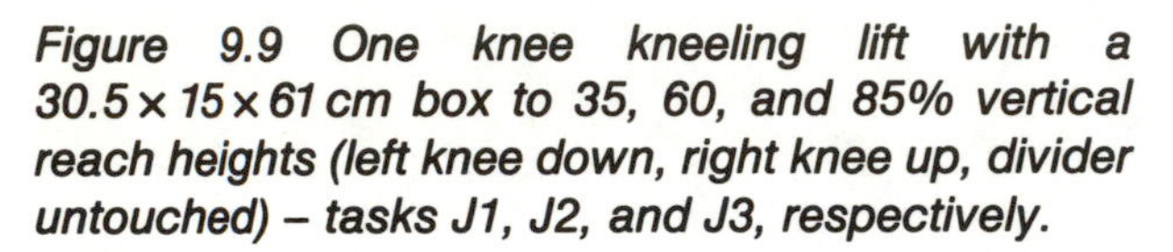

Figure 9.9 One knee kneeling lift with a 30.5 × 15 × 61 cm box to 35, 60, and 85% vertical reach heights (left knee down, right knee up, divider untouched) – tasks J1, J2, and J3, respectively.

Figure 9.10 Two knees kneeling lift with a 30.5 × 15 × 61 cm box to 35, 60, and 85% vertical reach heights (both knees down, divider untouched) – tasks K1, K2, and K3, respectively.

Figure 9.11 Standing lift with 15 × 61 × 30.5 cm box to 35, 60, and 85% vertical reach heights (knees bent, back straight, divider untouched) – tasks M1, M2, and M3, respectively.

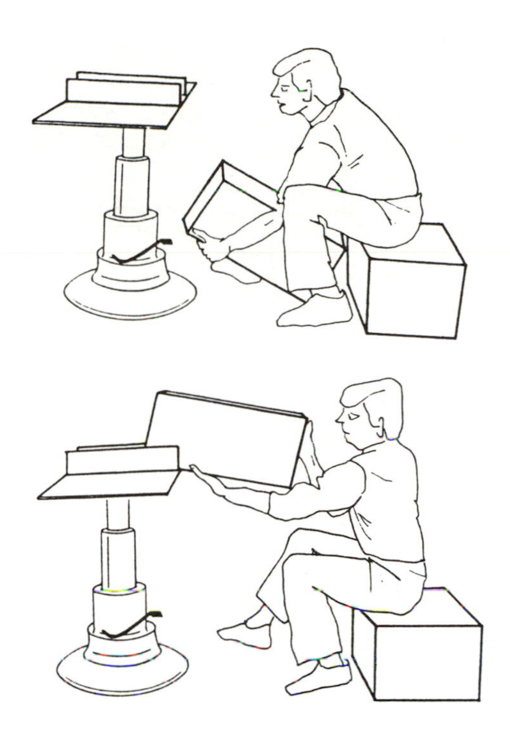

Figure 9.12 Sitting lift with a 15 × 61 × 30.5 cm box, sitting on a 30.5 cm high chair, to 35, 60, and 85% vertical reach heights (divider untouched) – tasks N1, N2, and N3, respectively.

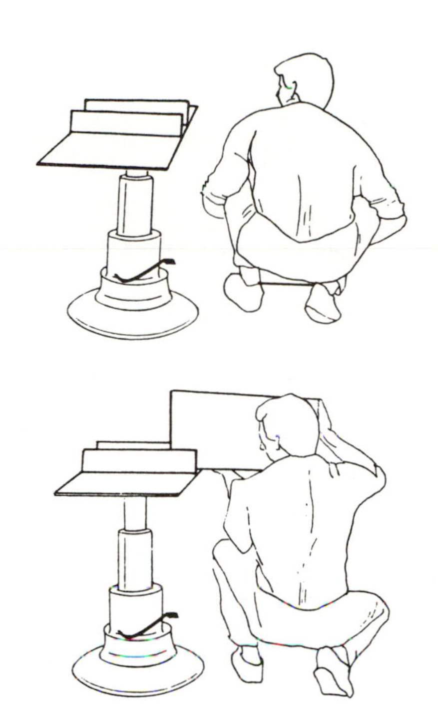

Figure 9.13 Squatting lift with a 15 × 61 × 30.5 cm box to 35, 60, and 85% vertical reach heights (divider untouched) – tasks P1, P2, and P3, respectively.

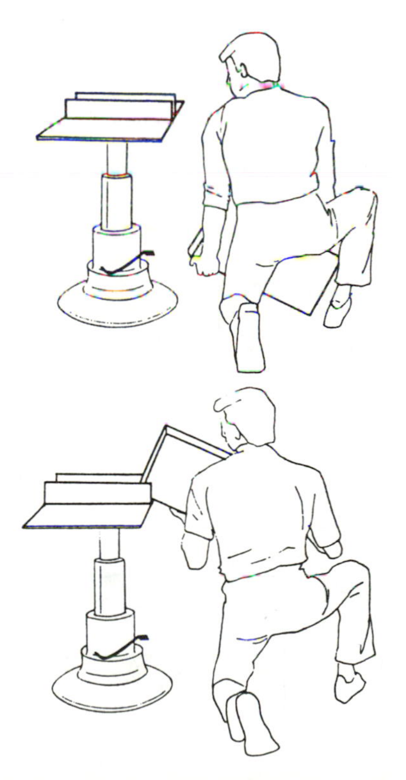

Figure 9.14 One knee kneeling lift with a 15 × 61 × 30.5 cm box to 35, 60, and 85% vertical reach heights (left knee down, right knee up, divider untouched) – tasks Q1, Q2, and Q3, respectively.

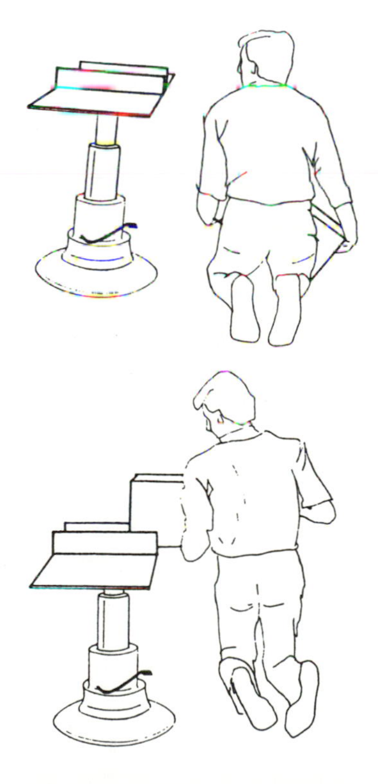

Figure 9.15 Two knees kneeling lift with a 15 × 61 × 30.5 cm box to 35, 60, and 85% vertical reach heights (both knees down, divider untouched) – tasks R1, R2, and R3, respectively.

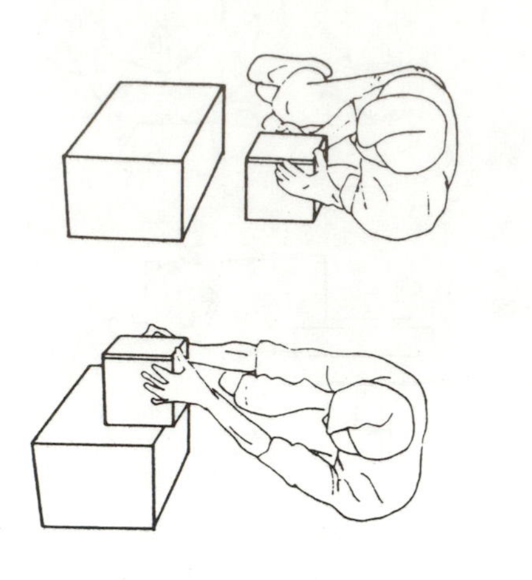

Figure 9.16 *Two hand side lift with legs at 90° with box in close position (operator lying on his/her side, legs curled to 90°, box 20 × 20 × 22 cm) – task SL2C1.*

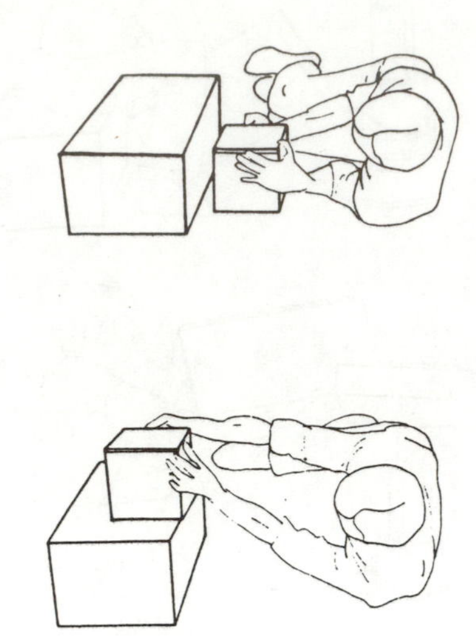

Figure 9.17 *Two hand side lift with legs at 90° with box at elbow distance (operator lying on his/her side, legs curled to 90°, box 20 × 20 × 22 cm) – task SL2F1.*

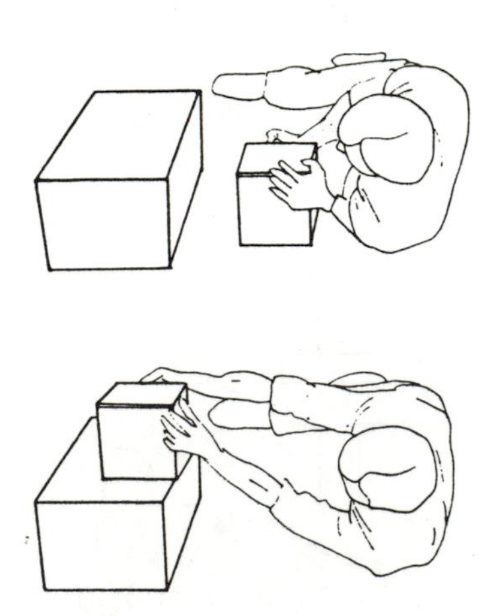

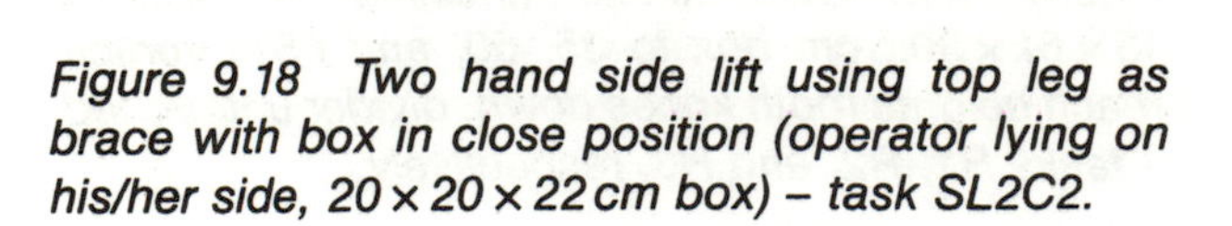

Figure 9.18 *Two hand side lift using top leg as brace with box in close position (operator lying on his/her side, 20 × 20 × 22 cm box) – task SL2C2.*

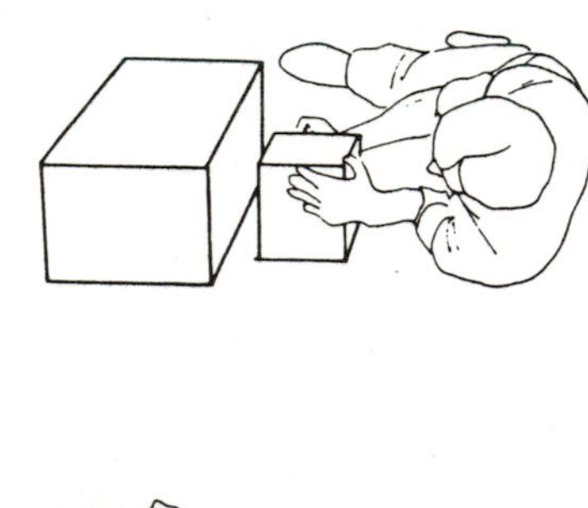

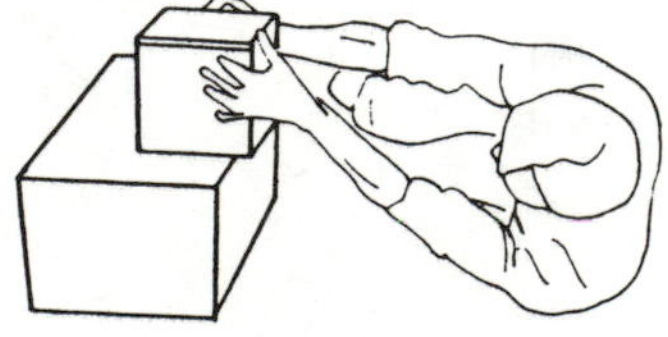

Figure 9.19 *Two hand side lift using top leg as brace with box at elbow distance (operator lying on his/her side, 20 × 20 × 22 cm box) – task SL2F2.*

Figure 9.20 One hand (right hand) lift while standing with a 25 × 25 × 28 cm box to 35, 60, and 85% vertical reach minus 25 cm heights (knees bent, back straight, divider untouched) – tasks S1, S2, and S3, respectively.

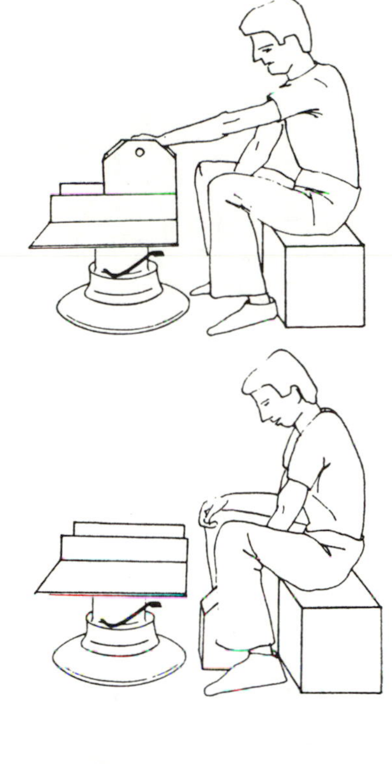

Figure 9.21 One hand (right hand) lift, sitting on a 30.5 cm high chair, with 25 × 25 × 28 cm box to 35, 60, and 85% vertical reach minus 25 cm heights (knees bent, back straight, divider untouched) – tasks T1, T2, and T3, respectively.

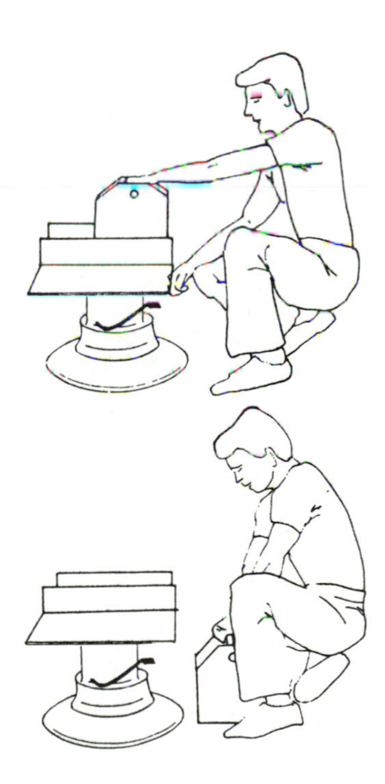

Figure 9.22 One hand (right hand) squatting lift with a 25 × 25 × 28 cm box to 35, 60, and 85% vertical reach minus 25 cm heights (knees bent, back straight, divider untouched) – tasks U1, U2, and U3, respectively.

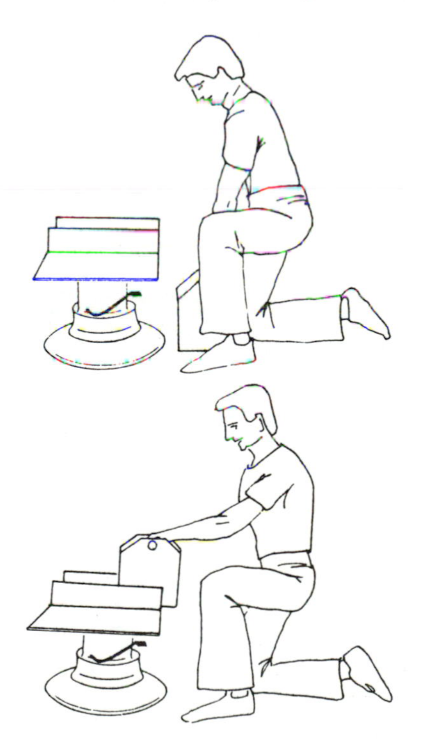

Figure 9.23 One hand (right hand) kneeling lift on one knee with a 25 × 25 × 28 cm box to 35, 60, and 80% vertical reach minus 25 cm heights (back straight, divider untouched) – tasks V1, V2, and V3, respectively.

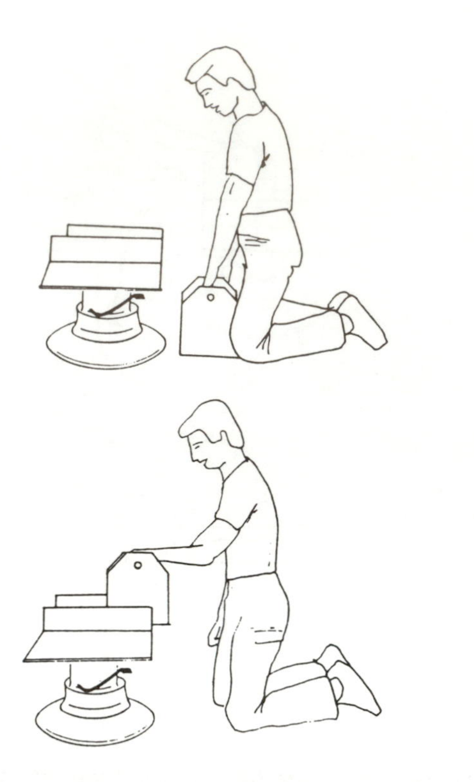

Figure 9.24 One hand (right hand) two knees kneeling lift with a 25 × 25 × 28 cm box to 35, 60, and 85% vertical reach minus 25 cm heights (divider untouched) – tasks W1, W2, and W3, respectively.

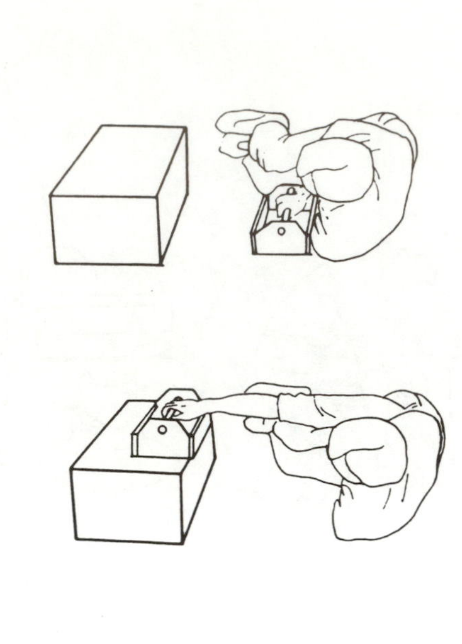

Figure 9.25 One hand (right hand) side lift with legs at 90° with a 37 × 20 × 12 cm box close to the body (operator lying on his/her side, legs curled to 90°) – task SL1C1.

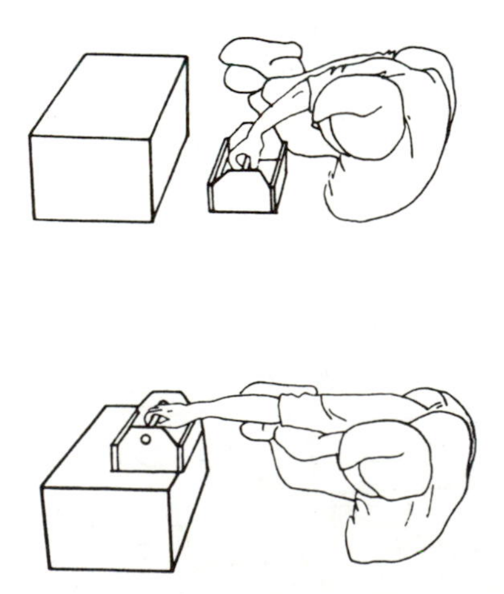

Figure 9.26 One hand side lift with legs at 90° with a 37 × 20 × 12 cm box at elbow distance (operator lying on his/her side, legs curled to 90°) – task SL1F1.

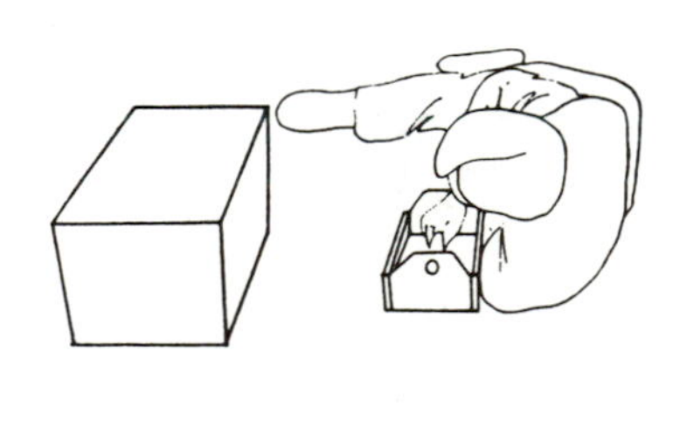

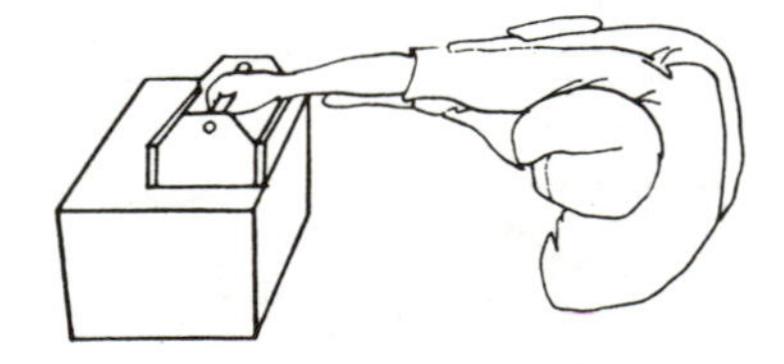

Figure 9.27 One hand (right hand) side lift using top leg as a brace with a 37 × 20 × 12 cm box in a close position (operator lying on his/her side) – task SL1C2.

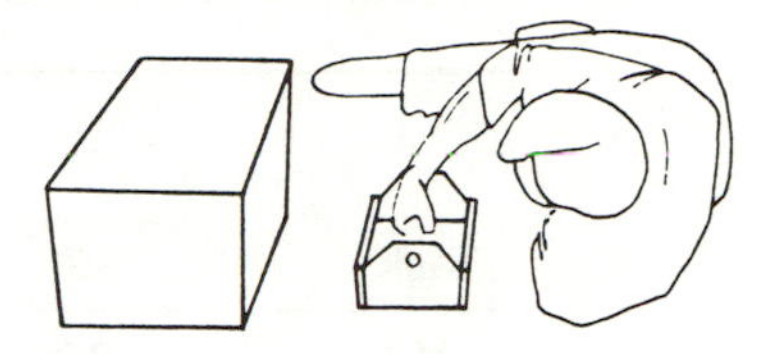

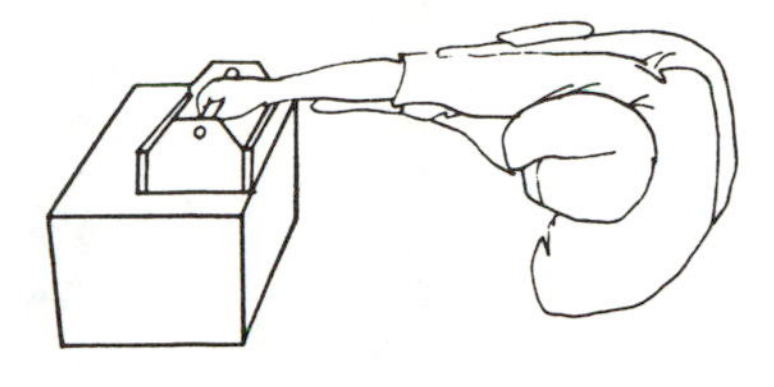

Figure 9.28 One hand (right hand) side lift using top leg as a brace with a 37 x 20 x 12 cm box at elbow distance (operator lying on his/her side) – task SL1F2.

Table 9.1 Recommended weight of lift (kg) for male industrial workers in unusual postures (figures 9.1–9.28)

Task	Population percentile				
	90	75	50	25	10
A1	27[a]	27[a]	27[a]	27[a]	27[a]
A2	27[a]	27[a]	27[a]	27[a]	27[a]
A3	25	27[a]	27[a]	27[a]	27[a]
B1	27[a]	27[a]	27[a]	27[a]	27[a]
B2	27	27[a]	27[a]	27[a]	27[a]
B3	21	24.5	27[a]	27[a]	27[a]
C1	27[a]	27[a]	27[a]	27[a]	27[a]
C2	23	26	27[a]	27[a]	27[a]
C3	18	21	24	27	27[a]
D1	27[a]	27[a]	27[a]	27[a]	27[a]
D2	26.5	27[a]	27[a]	27[a]	27[a]
D3	23	26.5	27[a]	27[a]	27[a]
E1	27[a]	27[a]	27[a]	27[a]	27[a]
E2	27[a]	27[a]	27[a]	27[a]	27[a]
E3	23	26	27[a]	27[a]	27[a]
F1	27[a]	27[a]	27[a]	27[a]	27[a]
F2	27	27[a]	27[a]	27[a]	27[a]
F3	21.5	25	27[a]	27[a]	27[a]
G1	27[a]	27[a]	27[a]	27[a]	27[a]
G2	24.5	27[a]	27[a]	27[a]	27[a]
G3	17	20	23	26	27[a]
H1	26.5	27[a]	27[a]	27[a]	27[a]
H2	21.5	25	27[a]	27[a]	27[a]
H3	15	18	21.5	25	27[a]
J1	27[a]	27[a]	27[a]	27[a]	27[a]
J2	26	27[a]	27[a]	27[a]	27[a]
J3	19.5	22	25	27[a]	27[a]
K1	27[a]	27[a]	27[a]	27[a]	27[a]
K2	26	27[a]	27[a]	27[a]	27[a]
K3	20	22	25	27[a]	27[a]
M1	27[a]	27[a]	27[a]	27[a]	27[a]
M2	27[a]	27[a]	27[a]	27[a]	27[a]

Table 9.1 Recommended weight of lift (kg) for male industrial workers in unusual postures (figures 9.1–9.28) (Continued)

Task	Population percentile				
	90	75	50	25	10
M3	27[a]	27[a]	27[a]	27[a]	27[a]
N1	27[a]	27[a]	27[a]	27[a]	27[a]
N2	26	27[a]	27[a]	27[a]	27[a]
N3	21	24	27	27[a]	27[a]
P1	27[a]	27[a]	27[a]	27[a]	27[a]
P2	25.5	27[a]	27[a]	27[a]	27[a]
P3	19	22	25.5	27[a]	27[a]
Q1	27[a]	27[a]	27[a]	27[a]	27[a]
Q2	27[a]	27[a]	27[a]	27[a]	27[a]
Q3	24	27	27[a]	27[a]	27[a]
R1	27[a]	27[a]	27[a]	27[a]	27[a]
R2	26.5	27[a]	27[a]	27[a]	27[a]
R3	23	26.5	27[a]	27[a]	27[a]
S1	9[b]	9[b]	9[b]	9[b]	9[b]
S2	9[b]	9[b]	9[b]	9[b]	9[b]
S3	9[b]	9[b]	9[b]	9[b]	9[b]
T1	9[b]	9[b]	9[b]	9[b]	9[b]
T2	9[b]	9[b]	9[b]	9[b]	9[b]
T3	9[b]	9[b]	9[b]	9[b]	9[b]
U1	9[b]	9[b]	9[b]	9[b]	9[b]
U2	9[b]	9[b]	9[b]	9[b]	9[b]
U3	9[b]	9[b]	9[b]	9[b]	9[b]
V1	9[b]	9[b]	9[b]	9[b]	9[b]
V2	9[b]	9[b]	9[b]	9[b]	9[b]
V3	9[b]	9[b]	9[b]	9[b]	9[b]
W1	9[b]	9[b]	9[b]	9[b]	9[b]
W2	9[b]	9[b]	9[b]	9[b]	9[b]
W3	9[b]	9[b]	9[b]	9[b]	9[b]
SL1C1	9[b]	9[b]	9[b]	9[b]	9[b]
SL1C2	9[b]	9[b]	9[b]	9[b]	9[b]
SL1F1	9[b]	9[b]	9[b]	9[b]	9[b]
SL1F2	9[b]	9[b]	9[b]	9[b]	9[b]
SL2C1	14	16	18	20	22
SL2C2	13	15.5	18	20.5	23
SL2F1	15	17	19	21	23
SL2F2	14.5	16.5	19	21.5	23.5

[a]Weight limited by the biomechanical design criterion (3930 N spinal compression).
[b]Weight limited by the IAP design criterion (90 mmHg).

Table 9.2 Recommended weight of lift (kg) for female industrial workers in unusual postures (figures 9.1–9.28)

Task	Population percentile				
	90	75	50	25	10
A1	18.5	20[a]	20[a]	20[a]	20[a]
A2	16.5	19	20[a]	20[a]	20[a]
A3	13	15	16.5	18	20
B1	17	19.5	20[a]	20[a]	20[a]
B2	14	16	18	20	20[a]
B3	11	12.5	14.5	16.5	18
C1	15.5	17.5	19.5	20[a]	20[a]
C2	12.5	14	16	18	19.5
C3	9	11	13	15	17
D1	18	20[a]	20[a]	20[a]	20[a]
D2	15	17	19	20[a]	20[a]

Table 9.2 Recommended weight of lift (kg) for female industrial workers in unusual postures (figures 9.1–9.28) (Continued)

Task	Population percentile				
	90	75	50	25	10
D3	11.5	13	15	17	18.5
E1	17	20[a]	20[a]	20[a]	20[a]
E2	14	16	18	20	20[a]
E3	12	13	15	17	18
F1	19	20[a]	20[a]	20[a]	20[a]
F2	15	17	19	20[a]	20[a]
F3	11.5	13	14.5	16	17.5
G1	16	18.5	20[a]	20[a]	20[a]
G2	13.5	15	16.5	18	19.5
G3	9	11	12	13	15
H1	14.5	16	18.5	20[a]	20[a]
H2	12	13.5	15	16.5	18
H3	8	9.5	11.5	13.5	15
J1	17	19	20[a]	20[a]	20[a]
J2	14	15.5	17	18.5	20
J3	11	12	13	14	15
K1	16	18.5	20[a]	20[a]	20[a]
K2	13.5	15	16.5	18	19.5
K3	10.5	11.5	13	14.5	15.5
M1	19.5	20[a]	20[a]	20[a]	20[a]
M2	16.5	18.5	20[a]	20[a]	20[a]
M3	13.5	15	16.5	18	19.5
N1	16	18.5	20[a]	20[a]	20[a]
N2	14	15.5	17.5	19.5	20[a]
N3	11	12.5	14	15.5	17
P1	15.5	17.5	20	20[a]	20[a]
P2	13.5	15	17	19	20[a]
P3	10.5	12	13.5	15	16.5
Q1	17.5	20[a]	20[a]	20[a]	20[a]
Q2	15.5	17	19	20[a]	20[a]
Q3	13	14.5	15.5	16.5	18
R1	19	20[a]	20[a]	20[a]	20[a]
R2	16	17.5	19	20[a]	20[a]
R3	13	14.5	16	17.5	19
S1	6[b]	6[b]	6[b]	6[b]	6[b]
S2	6[b]	6[b]	6[b]	6[b]	6[b]
S3	6[b]	6[b]	6[b]	6[b]	6[b]
T1	6[b]	6[b]	6[b]	6[b]	6[b]
T2	6[b]	6[b]	6[b]	6[b]	6[b]
T3	6[b]	6[b]	6[b]	6[b]	6[b]
U1	6[b]	6[b]	6[b]	6[b]	6[b]
U2	6[b]	6[b]	6[b]	6[b]	6[b]
U3	6[b]	6[b]	6[b]	6[b]	6[b]
V1	6[b]	6[b]	6[b]	6[b]	6[b]
V2	6[b]	6[b]	6[b]	6[b]	6[b]
V3	6[b]	6[b]	6[b]	6[b]	6[b]
W1	6[b]	6[b]	6[b]	6[b]	6[b]
W2	6[b]	6[b]	6[b]	6[b]	6[b]
W3	6[b]	6[b]	6[b]	6[b]	6[b]
SL1C1	5.5	6	6[b]	6[b]	6[b]
SL1C2	5.5	6	6[b]	6[b]	6[b]
SL1F1	5.5	6	6[b]	6[b]	6[b]
SL1F2	5.5	6	6[b]	6[b]	6[b]
SL2C1	6	7.5	9	10.5	12
SL2C2	6.5	8	9	10	11.5
SL2F1	7	8	9.5	11	12
SL2F2	7	8	9.5	11	12

[a]Weight limited by the biomechanical design criterion (2689 N spinal compression).
[b]Weight limited by the IAP design criterion (90 mmHg).

Pushing

The isometric (static) pushing force data for postures are shown in Figures 9.29–9.37. The design data for these postures are given in Tables 9.3–9.9 (figures and data modified from Kroemer, 1969).

Figure 9.29 Pushing with the back, squatting.

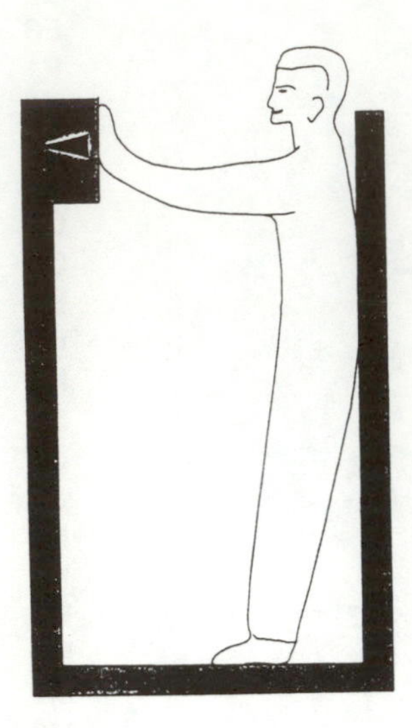

Figure 9.30 Pushing with one or two hands, standing.

Figure 9.31 Pushing while standing, arms extended.

Figure 9.32 Pushing with both hands, operator bracing one leg.

Figure 9.33 Pushing with shoulder, standing with one foot braced against a footrest.

Figure 9.34 Pushing with both hands, standing with one foot braced against a footrest.

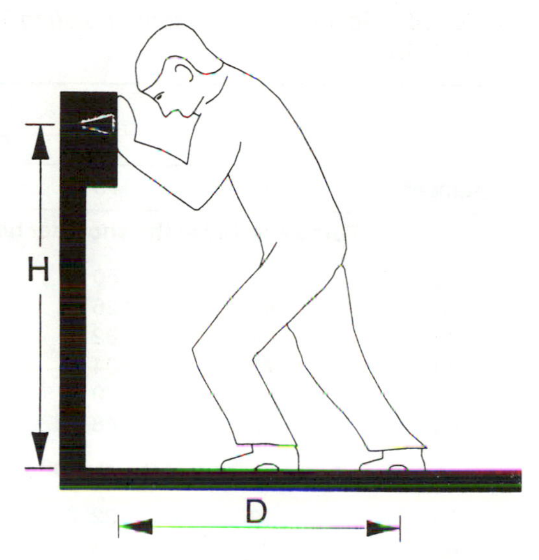

Figure 9.35 Pushing with both hands, standing, no anchor for foot support (operator selecting foot distance and height of force application).

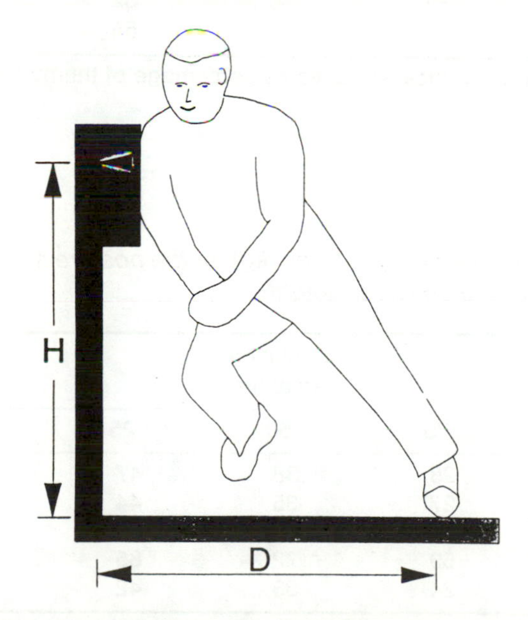

Figure 9.36 Pushing with shoulders, standing, no footrest or brace for anchoring (operator selecting foot distance and height of force application).

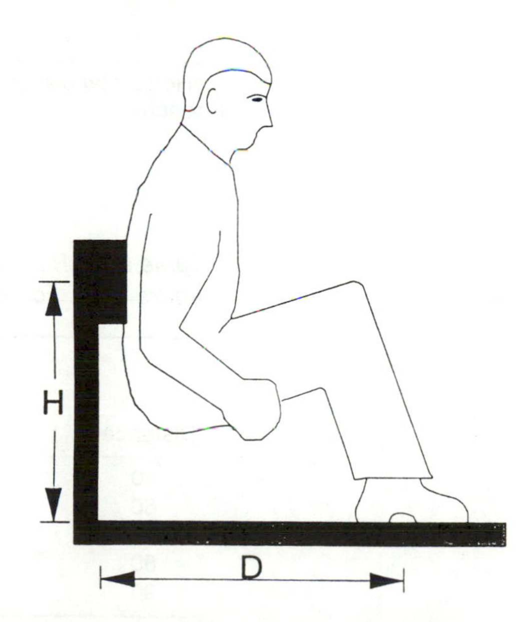

Figure 9.37 Pushing with back, squatting, no brace or footrest for anchoring (operator selecting foot distance and height of force application).

Table 9.3 Recommended isometric push force (kg) for the posture shown in figure 9.29. Force exertion at 40% of shoulder height

	Population percentile				
Distance[a]	90	75	50	25	10
80	98	143	194	245	294
90	95	128	165	202	236
100	103	134	169	203	234
110	123	158	197	236	271
120	122	159	200	241	278
130	113	145	180	216	248

[a]Horizontal distance between two vertical surfaces as a percentage of thumb-tip reach.

Table 9.4 Recommended isometric push force (kg) for the posture shown in figure 9.30.

	Population percentile				
Distance[a]	90	75	50	25	10
Force exerted at the shoulder height with both hands					
50	41	50	59	153	172
60	47	126	68	79	89
70	65	82	100	119	136
80	79	104	131	159	183
90	60	79	100	121	139
100	33	48	66	83	99
Force exerted at the shoulder height with one hand					
50	18	22	27	31	35
60	21	26	30	35	40
70	24	30	37	43	49
80	34	43	53	63	72
90	28	39	50	62	72
100	21	32	44	55	66

[a]Horizontal distance between the vertical surfaces as percentage of thumb-tip reach.

Table 9.5 Recommended isometric push force (kg) for the posture shown in figure 9.31. Force exerted at the shoulder height.

	Population percentile				
Distance[a]	90	75	50	25	10
50	20	28	38	47	56
60	19	27	35	44	52
70	32	42	53	64	75
80	47	59	72	85	97
90	16	24	33	42	51

[a]Percent of span (distance between a person's fingertips, arms extended to sides).

Table 9.6 Recommended isometric push force (kg) for the posture shown in figure 9.32.

		Population percentile				
Height[a]	Distance[b]	90	75	50	25	10
50	80	44	55	68	80	91
50	100	51	64	79	94	107
50	120	58	68	79	91	101
70	80	52	62	73	85	94
70	100	44	58	74	90	105
70	120	65	74	83	93	101
90	80	45	54	64	74	83
90	100	43	55	69	82	95
90	120	69	78	88	98	107

[a]Height of force application as percentage of shoulder height.
[b]Horizontal distance between vertical surfaces as percentage of shoulder height.

Table 9.7 Recommended isometric push force (kg) for the posture shown in figure 9.33.

		Population percentile				
Height[a]	Distance[b]	90	75	50	25	10
60	70	55	66	78	89	100
60	80	64	75	87	99	110
60	90	62	71	81	91	99
70	60	44	51	59	67	73
70	70	55	63	71	80	87
70	80	55	64	74	84	93
80	60	36	44	53	62	70
80	70	48	56	65	74	80
80	80	47	56	65	74	83

[a]Height of force application as a percentage of shoulder height.
[b]Horizontal distance between the vertical surfaces as percentage of shoulder height.

Table 9.8 Recommended isometric push force (kg) for the posture shown in figure 9.34.

		Population percentile				
Height[a]	Distance[b]	90	75	50	25	10
70	70	44	53	63	73	82
70	80	50	60	70	80	90
70	90	42	51	60	69	78
80	70	40	47	56	65	72
80	80	40	47	56	65	72
80	90	44	49	54	59	64
90	70	32	38	44	50	56
90	80	34	39	46	53	58
90	90	39	44	49	54	59

[a]Height of force application as a percentage of shoulder height.
[b]Horizontal distance between the vertical surfaces as a percentage of shoulder height.

Table 9.9 Recommended isometric push forces (kg) for postures shown in Figures 9.35–9.37.

Posture	Floor (u)[a]	Population percentile				
		90	75	50	25	10
Figure 9.35	Very Slippery ($u=0.3$)	15	17	20	23	25
Figure 9.36	$u=0.3$	15	17	20	23	25
Figure 9.37	$u=0.3$	14	17	19	21	24
Figure 9.35	Moderately Slippery ($u=0.6$)	22	26	31	36	40
Figure 9.36	$u=0.6$	23	27	32	37	41
Figure 9.37	$u=0.6$	24	28	33	38	42

[a] u, Coefficient of friction.

Pulling

The pulling design data in postures shown in Figures 9.35–9.37 are given in Table 9.10.

Table 9.10 Recommended minimum pulling force (kg) for the postures shown in Figures 9.35–9.37.

Force at least:	Applied with	Condition (u = coefficient of friction)
11	Both hands or one shoulder or the back	With low traction ($0.2 \leq u \leq 0.3$)
20	Both hands or one shoulder or the back	With medium traction ($u=0.6$)
30	Both hands or one shoulder or the back	With high traction ($u \geq 0.9$)
51	Both hands or one shoulder or the back	If braced against a vertical wall 50–150 cm from and parallel to the panel or if anchoring the feet on a perfectly non-slip ground (like a footrest)

References

Ayoub, M.M., Smith, J.L., Selan, J.L., Chen, H.C., Lee, Y.H., Kim, H.K. and Fernandez, J.E., 1987. Manual material handling in unusual postures. Technical Report, Department of Industrial Engineering, Texas Tech University, Lubbock, Texas, USA.

Kroemer, K.H.E. 1969. *Push Forces Exerted in 65 Common Working Positions.* Report no. AMRL-TR-68-143, Wright-Patterson Air Force Base, Ohio.

Chapter 10

Designing and evaluating multiple-task manual materials handling jobs: using the guide

Introduction

Matching the requirements of a job with the capabilities of a specific population, or designing jobs using the capability data of a certain population group, is the approach ergonomists and designers have adopted in order to reduce the risks and severity of musculoskeletal injuries resulting from handling materials manually. A number of job design and employee screening procedures have been developed, based on either the psychophysical, the physiological, the biomechanical, or the epidemiological design criterion. Ayoub and Mital (1989) provide a detailed review of these procedures.

The majority of job design/redesign/analysis procedures, including the users' manual associated with the revised NIOSH guidelines (Waters *et al.*, 1993) are, however, for a specific MMH task (lifting, pushing, or carrying, for instance). Manual handling jobs which comprise a single task are a rarity in industry. In fact, most MMH jobs involve at least two different kinds of manual handling tasks; many are a combination of several different tasks. Therefore, a procedure is needed to analyse or design manual handling jobs that are a combination of several different tasks (lifting, lowering, pushing, carrying, etc.).

The concept of a procedure that can assist in the design and analysis of a multiple-task MMH job was first proposed by Mital (1983). This procedure was subsequently revised (Jiang and Mital, 1986; Mital, 1991 and 1995). The last revision (Mital, 1995) included a brief description of how this guide may be used in designing/analysing multiple-task MMH jobs. This procedure is described in this chapter, and three case studies are solved to demonstrate the use of the *Guide*.

Design/evaluation procedure

The generalized manual materials handling job design/evaluation procedure is based on the concept that a person's ability to perform a manual materials handling job is based on his/her capability to perform individual tasks that make up that job. The main steps of the design/evaluation procedure are:

1. Break the manual materials handling job into individual manual materials handling tasks (lifting, lowering, pushing, pulling, and carrying). There may be more than one task of each type. Prepare workplace layout showing appropriate distances, record total working duration including break (lunch, coffee, etc.) time, and record the cycle time or other appropriate measure for computing handling frequency.
2. Select the population percentile for which the job is to be designed/evaluated. One may choose, for example, the upper 75% or 90% male or female population or any other population

percent. If a mixed population is chosen, for instance the upper 75% male and the upper 90% female, analysis must be performed twice (once for males and once for females) and the job design/evaluation must be based on the higher of the two risk potential values (see step 4 for the determination of the risk potential).

3. For each MMH task, determine the recommended work rate. Use tables in chapters 4–9 to obtain values of recommended weight/force for given frequency and distance moved. Distance data should be available from either the actual workplace layout or the proposed workstation design plans. Frequency data should be available from production requirements. Use these work rate values if designing new jobs; otherwise go to step 4.
4. Compare recommended work rate with actual work rate by calculating the risk potential R (= actual work rate / recommended work rate). Note that calculation of work rate is not essential for calculating R, as R is also the same as the ratio of actual weight (or force) to recommended weight (or force). Work rate calculations help in obtaining alternative solutions.
5. If, for any task, $R > 1$ then redesign that task; otherwise accept the job. (R can be reduced to 1 by reducing either the weight/force or the distance moved/covered or the frequency of handling. This provides several alternative solutions. Obviously, the least expensive solution should be implemented.)

The application of the above step-by-step procedure and the use of this guide are demonstrated in three case studies.

Case study 1

The stacking and palletizing job in a cattle feed bagging plant is analysed. The job involves stacking feed bags (22.73 kg each) on a wooden pallet (20.42 cm high) located approximately 1.83 m from the conveyor (Figure 10.1).

A male operator lifts the bag off the conveyor, carries it to the pallet, stacks it, and then returns for the next bag (the job only employs male workers). Forty-five bags (5 per layer, total 9 layers) are stacked per pallet. Bags are 15.24 cm thick, 40.64 cm long (dimension in the sagittal plane – dimension away from the body), and 76.20 cm wide (dimension in the frontal plane which divides the body into two halves, front and

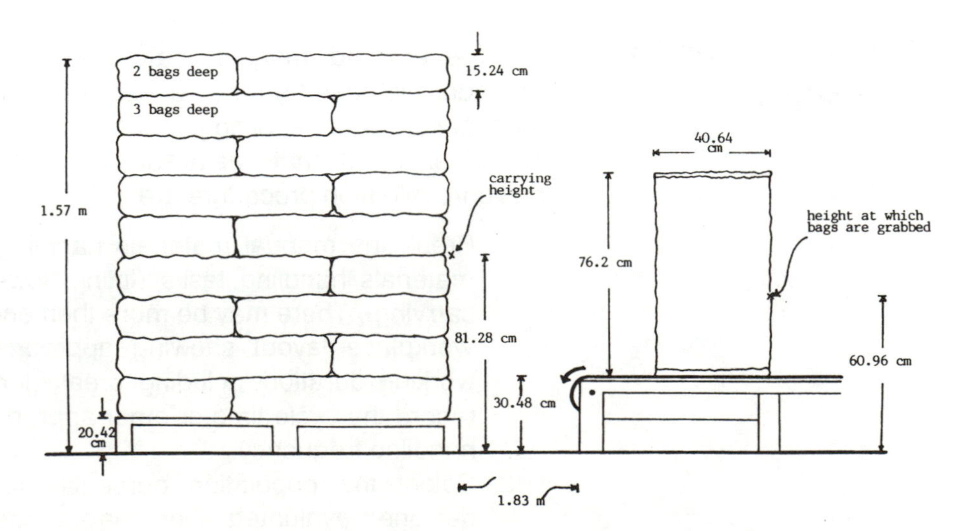

Figure 10.1 Workplace layout (Case study 1 – Mital, 1991).

back). While carrying, the effective bag dimension in the sagittal plane is reduced on account of deformation to 38.10 cm. Bags are carried a distance of 1.83 m at a height of 81.28 cm (Figure 10.1). A time study indicated that, on the average, it takes 575 s to completely load the pallet. The worker operates a 600 min (10 h shift) each day including 60 min for different breaks.

Complaints of persistent arm, shoulder, and back pain warranted evaluation of the job. The step-by-step procedure, described above, was used. The goal was that the job should be acceptable to the upper 75% of the male population.

A review of the job showed that in order to load the first 4 layers, bags must be lowered onto the pallet; for the subsequent 5 layers, bags must be lifted. The lowering and lifting distances will change with each layer. From the cycle time (575 s) handling frequency was obtained as follows: 45 bags are loaded in 9.58 min (= 575 s/60 s) leading to a frequency of 4.69 bags/min (= 45 bags/9.58 min). The total working duration (shift duration) was 10 hours (600 min) including formal breaks.

The job was broken into 11 MMH tasks (6 lifting tasks, 4 lowering tasks, and 1 carrying task). The sequence of these tasks was as follows:

Task 1 (lifting) – lifting bags from the conveyor
Task 2 (carrying) – carrying bags from the conveyor to the pallet
Task 3 (lowering) – lowering bags to the first layer height
Task 4 (lowering) – lowering bags to the second layer height
Task 5 (lowering) – lowering bags to the third layer height
Task 6 (lower) – lowering bags to the fourth layer height
Task 7 (lifting) – lifting bags to the fifth layer height
Task 8 (lifting) – lifting bags to the sixth layer height
Task 9 (lifting) – lifting bags to the seventh layer height
Task 10 (lifting) – lifting bags to the eighth layer height
Task 11 (lifting) – lifting bags to the ninth layer height

For each task actual and recommended work rates and risk potential were calculated. All distances moved (horizontal and vertical) were centre-to-centre distances.

Table 10.1 shows the input values, actual and recommended work rates, and risk potential for all 11 tasks. A detailed calculation for Task 1 (lifting) is given below:

Frequency of lifting = 4.69/minute
Box (bag) size = 38.1 cm
Lifting starting point (x) = 60.96 cm
Lifting ending point (y) = 81.28 cm
Vertical height lifted ($y - x$) = 81.28 — 60.96 = 20.32 cm
Height range of lifting = floor to 80 cm
Working duration = 10 hours
Population percentile = 75th male
Actual weight of the bag = 22.73 kg

From the above data, actual work rate was determined as follows:

Actual work rate (kg-m/min) = bag weight × distance lifted × lifting frequency
= 22.73 × 0.2032 × 4.69
= 21.66 kg-m/min

In order to determine the recommended work rate, first the recommended weight of lift needs to be determined. Table 4.2 of the

Table 10.1. Actual and recommended work rates (kg-m/min) and risk potential for various tasks of case study 1

Task	Input data	Actual work rate (*A*)	Recommended work rate (*B*)	Risk potential $R^a = A/B$
1. Lifting from conveyor	Box size = 38.1 cm Vertical height = 20.32 cm Frequency = 4.69/min Height range = F–80 cm	*W* = 22.73 × 0.2032 × 4.69 = 21.66	*W* = 15.11 × 0.2032 × 4.69 = 13.53	= 21.66/13.41 = 1.61
2. Carrying from conveyor to pallet	Frequency = 4.69/min Carrying distance = 1.83 m Carrying height = 0.81 m	W = 22.73 × 1.83 × 4.69 = 195	W = 20.34 × 1.83 × 4.69 = 175	= 195/175 = 1.11
3. 1st layer lower	Vertical height = 53.35 cm Rest of the input = same as in Task 1	*W* = 22.73 × 0.5335 × 4.69 = 56.87	*W* = 15.11 × 4.69 × 0.5335 = 37.80	= 56.87/37.80 = 1.50
4. 2nd layer lower	Vertical height = 38.1 cm Rest of the input = same as in Task 1	*W* = 22.73 × 0.381 × 4.69 = 40.61	*W* = 15.11 × 4.69 × = 26.99	= 40.61/26.99 = 1.50
5. 3rd layer lower	Vertical height = 22.86 cm Rest of the input = same as in Task 1	*W* = 22.73 × 0.2286 × 4.69 = 24.37	*W* = 15.11 × 4.69 × 0.2286 = 16.19	= 24.37/16.19 = 1.50
6. 4th layer lower	Vertical height = 7.62 cm Rest of the input = same as in Task 1	*W* = 22.73 × 0.0762 × 4.69 = 8.12	*W* = 15.11 × 4.69 × 0.0762 = 5.39	= 8.12/5.39 = 1.50
7. 5th layer lift	Box size = 38.1 cm Vertical height = 7.62 cm Frequency = 4.69/min Height range =F–80 cm	*W* = 8.12	*W* = 5.39	= 1.50
8. 6th layer lift	Box size = 38.1 cm Vertical height = 22.86 cm Frequency = 4.69/min Height range = 80–132 cm	*W* = 24.37	*W* = 15.03 × 4.69 × 0.2286 = 16.11	= 24.37/16.11 = 1.51
9. 7th layer lift	Vertical height = 38.1 cm Rest of the input = same as in Task 8	*W* = 40.61	*W* = 15.03 × 4.69 × 0.381 = 26.85	= 40.61/26.85 = 1.51
10. 8th layer lift	Vertical height = 53.35 cm Rest of the input = same as in Task 8	*W* = 56.87	*W* = 15.03 × 4.69 × 0.5335 = 37.60	= 56.87/37.60 = 1.51
11. 9th layer lift	Vertical height = 68.58 cm Height range = 80–183 cm Rest of the input = same as in Task 8	*W* = 22.73 × 0.6858 × 4.69 = 73.1	*W* = 15.03 × 4.69 × 0.6858 = 48.34	= 73.1/48.34 = 1.51

[a]*R* is also the same as the ratio of actual weight (force)/recommended weight (force). Calculation of work rate helps in identifying alternative solutions.

guide provides recommended weights for males as a function of lifting frequency, box size, and lifting height range. For Task 1, the lifting height range is from the floor to 80 cm (approximately), the lifting frequency is 4.69 bags per minute, and the box size (effective bag size) is 38.1 cm. Using the floor to 80 cm height region of Table 4.2 and the 75% capable male values, we first interpolate between the lifting frequencies of 4/min and 8/min for the frequency of 4.69/minute and then between the box sizes of 34 cm and 49 cm for the box size of 38.1 cm.

The first interpolation, for box size of 34 cm, is between 17 kg and 10 kg weight. This value is determined to be 15.79 kg. The second interpolation, for box size 49 cm, is between 14 and 10 kg. This value is determined to be 13.31 kg. Now we interpolate between these two values for the box size of 38.1 cm. The recommended weight value for a box size of 38.1 cm and lifting frequency of 4.69/minute is determined to be 15.11 kg. Note that this weight is for a working duration of 8 hours. Since the actual working duration is 10 hours, this weight must be reduced as per Table 4.4. The duration multiplier obtained from Table 4.4 is 0.932. The recommended weight for Task 1, thus, is 15.11 × 0.932 or 14.08 kg and the recommended work rate is 14.08 kg × 4.69 lifts/min × 0.2032 m lifting height or 13.41 kg-m/min. The risk potential, *R*, for Task 1 is 1.61 (=21.66/13.41). In a similar manner, the actual work rate, the recommended work rate, and risk potential for all other tasks may be calculated.

Since all eleven tasks have an *R* value greater than 1, the job must be redesigned by reducing either the weight or the distance travelled or the frequency. Between the weight and frequency, it is more important to reduce the frequency. The economic impact of each alternative must also be considered prior to implementing the solution. For instance, it is not simple to implement a reduction in the weight of the bag. Such a solution, besides resulting in the expense of a new bag, would also increase the overall frequency for the fixed tonnage that must be handled daily. Reduction in frequency, similarly, will reduce production capacity and, perhaps, result in an inability to meet the market demands. Solutions such as changing the bag dimensions or adding a relief worker to reduce the overall frequency should be considered.

Case study 2

A major producer of athletic goods receives unwashed knitted material that must be dyed, processed, and then folded into bundles by a machine called a calender. These bundles weigh about 20.45 kg and are 38.1 cm long in the sagittal plane. A worker removes each bundle from the calender, which is about 1.35 m from the floor, and lowers it to a table 63.5 cm high, and ties it. The worker then carries the bundle 6.1 m (at his knuckle height – approximately 79 cm) to a scale, weighs it, and then tags the bundle with a computer-generated card. Next, the bundle is loaded on a buggy for temporary storage (2.6 m away from the scale), from whence it is trucked to another plant for sewing. The worker is on the job 7.5 h each day. In addition, he gets a 30 min break for lunch. The plant processes 470 bundles for each calender per day. Figure 10.2 shows the workplace layout. As the workplace layout shows, the worker must turn 90° on several occasions: while lowering the bundle from the calender, while lifting and carrying the bundle to the scale, and while lifting and carrying it to the buggy.

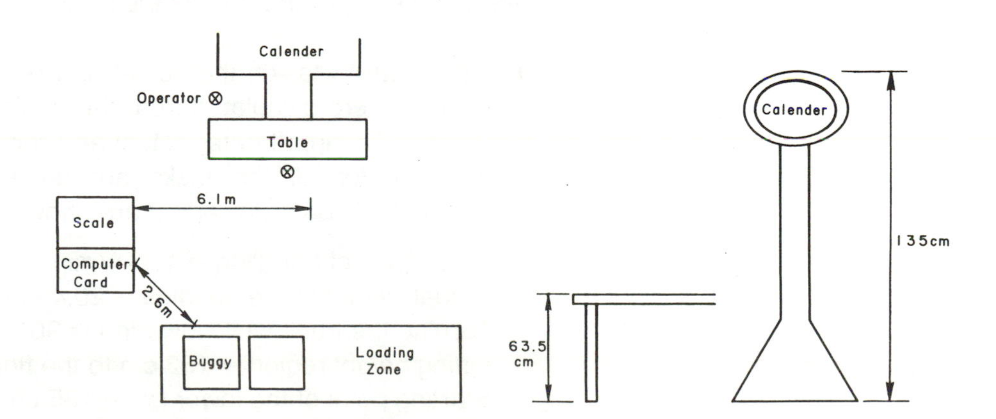

Figure 10.2 Schematic workplace layout (Case study 2 – Jiang and Mital, 1986).

Table 10.2 Actual and recommended work rates (kg-m/min) and risk potentials for various tasks of case study 2

Task	Input data	Actual work rate (*A*)	Recommended work rate (*B*)	Risk potential (= *A/B*)
1. Lowering bundles from calender to table	Box size = 38.1 cm Turning = 90° Vertical height = 71.5 cm Height range = 132 cm–F Frequency = 1.044/min	= 20.45 × 0.715 × 1.044 =15.26	= 17.05 × 0.715 × 1.044 = 12.72	= 15.26/12.72 = 1.2
2. Tying bundles				
3. Lifting bundles	Vertical height = 15.5 cm Height range = F–80 cm Rest of the input = same as Task 1	= 20.45 × 0.155 × 1.044 = 3.31	= 17.96 × 0.155 × 1.044 = 2.90	= 3.31/2.90 = 1.14
4. Carrying bundles to scale	Carrying distance = 6.1 m Carrying height = 79 cm Frequency = 1.044/min	=20.45 × 6.1 × 1.044 = 130.23	= 24.5 × 6.1 × 1.044 = 156.02	= 130.23/156.02 = 0.83
5. Generating computer cards				
6. Lifting bundles	Same as Task 3	= 3.31	= 2.90	= 1.14
7. Carrying bundles	Carrying distance = 2.6 m Rest of the input = same as Task 4	= 20.45 × 2.6 × 1.044 = 55.51	= 27 × 2.6 × 1.044 = 73.29	= 55.51/73.29 = 0.76

The plant manager wants to redesign this job to accommodate 75% of the male population. From the production data, the handling frequency is calculated to be 1.044 bundles per minute (= 470 bundles/450 min).

This job was divided into 7 tasks; only 5 of these involve manual handling. These tasks are:

Task 1 (lowering) – lowering bundles from the calender to the table (turning 90°)
Task 2 – Tying the bundle
Task 3 (lifting) – lifting bundles from the table (turning 90°)
Task 4 (carrying) – carrying bundles to the scale for weighing
Task 5 – generating computer card
Task 6 (lifting) – lifting bundles from the scale (turning 90°)
Task 7 (carrying) – carrying bundles to the buggy

There is lowering at the end as the bundles are simply dropped into the buggy.

For the 5 MMH tasks, the actual and recommended work rates and risk potential were calculated from the design data given in Tables 4.2, 4.6, and 7.1. The input data, actual and recommended work rates, and risk potential for all the tasks are shown in Table 10.2. Sample calculations for Task 1 (lowering) are shown below:

Frequency of handling = 1.044/min
Actual weight of the bundles = 20.45 kg
Turning the trunk while lowering = 90°
Lifting height region = 132 cm to the floor (approximately)
Starting point of the lower (x) = 135 cm
Ending point of the lower (y) = 63.5 cm
Lowering height ($x - y$) = 71.5 cm
Box (bundle) size = 38.1 cm

Population percentile = male upper 75%
Working duration = 8 hours

From the input data, the actual work rate was calculated to be 15.26 kg-m/min (= 20.45 kg × 1.044 bundles/min × 0.715 m). The recommended weight, determined from Table 4.2 by interpolation, was 20.11 kg. This was multiplied by 0.848, the asymmetrical handling multiplier from Table 4.6. The final recommended weight thus was 17.05 kg and the recommended work rate was 12.72 kg-m/min (= 17.05 kg × 1.044 bundles/min × 0.715 m lowering height). This provided a risk potential value of $R = 1.2$ (= 15.26/12.72).

As seen in Table 10.2, all lifting and lowering tasks have an R value of greater than 1. The solution to the problem in this case involved a change in the layout of the workplace. The changes reduced the trunk turning to under 30°.

Case study 3

The job involves a device, known as an End-of-Train (EOT) Device, used by railroad companies. It is mounted on goods trains at the rear (101.6 cm above the ground) to indicate danger (red light – Figure 10.3).

The device weighs 25 kg, and is 60.96 cm, 30.48 cm wide (distance between hands), and 25.4 cm deep. The weight (battery) of the device is concentrated in the upper half (c.g. is offset by 15.24 cm). The bottom of the device has a 45.72 cm long vertical projection. There is a single vertical handle on the right side of the EOT Device.

The job involves a male worker removing the EOT Device from the goods train, lifting it and placing it on his right shoulder, turning 180°, walking 18.29 m across the tracks, lifting the EOT Device to 152.4 cm height, holding it and gradually sliding it (away from the body in the sagittal plane) on a narrow rack placed on the tailgate of a pickup truck. On an average, one train is handled every 30 min. This work continues all day. The job, thus, involves lifting, turning, carrying, holding, and sliding.

The railroad company desired that the job be analysed using the 75% males. For this purpose, the job was divided into three tasks:

Task 1 – lifting EOT Device from 101.6 cm height to 152.4 cm (shoulder height)
Task 2 – carrying the load 18.29 m
Task 3 – moving the load from the shoulder and sliding it on the rack.

Only the first two tasks could be analysed by the guide.

For the lifting task, height range was between 80 and 183 cm and the object size was 30.48 cm. The 75th percentile lifting weight limit (from Table 4.2) was determined to be 22 kg. To this number, the following corrections were applied:

asymmetrical lifting correction from Table 4.6 (0.848 for max. turning)
couplings correction from Table 4.8 (0.925 for poor quality)
load clearance correction from Table 4.9 (0.87 for 3 mm clearance)
load asymmetry correction from Table 4.7 (0.93 for 15.24 cm c.g. offset)

Using these correction factors, the recommended weight was determined to be:

$$22 \times 0.848 \times 0.925 \times 0.87 \times 0.93 = 13.96 \text{ kg}$$

Note that determination of work rate is not critical in this case, owing to a very small frequency of handling (0.033/min). The risk potential for the lifting task was 1.79 (= 25/13.96).

The recommended weight for carrying for a distance of 18.3 m was computed from the limit for 8.3 m distance (16.5 kg) by reducing the recommended weight limit by 1 kg for every 4 m distance. The final recommended weight was 14 kg. The risk potential for the carrying task was 1.78 (= 25/14).

It is worth noting that for both tasks the risk potential was about the same and weight is the most critical factor. The solution to this problem lies in redesigning the EOT unit or changing the way the job is done. Since it is unlikely that the railroad company would be able to find a lighter EOT device as it would need a much lighter battery, or new and

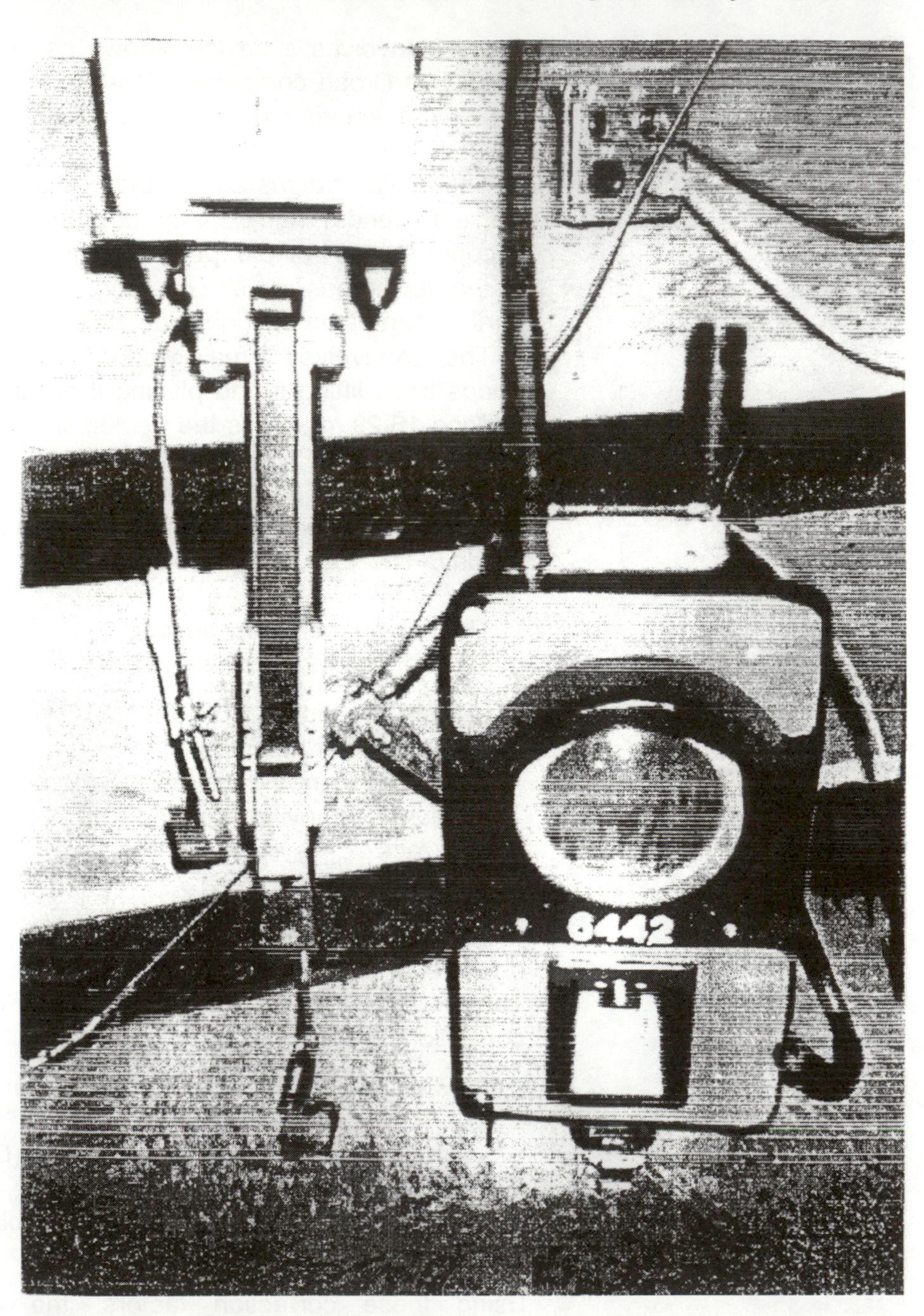

Figure 10.3 End-of-Train (EOT) device (not to scale; Case study 3 – Mital and Ramakrishnan, 1996).

expensive technology, the way the job is done must be changed. First, instead of lifting the EOT device onto the shoulder, it must be lowered on some sort of a trolley. This would also eliminate the carrying task. At the placement end, the EOT device could be mechanically lifted onto the truck tailgate. The trolley could be equipped with this mechanical device.

Summary

The cases discussed in this chapter are actual and are included here to demonstrate the use of this guide. As the discussion following each case suggests, there are a number of alternatives; not all are equally desirable when considering the costs involved. It should also be noted that the job design/evaluation procedure outlined here ensures that only those tasks for which the risk potential exceeds 1 need to be redesigned. This eliminates the need to redesign the whole job. For instance, in case 2, only the lifting tasks really needed to be redesigned.

One can write software to analyse or design multiple task MMH jobs based on the procedure given in this chapter. Such a program is simple to write and would require that the user include all the design tables given in Chapters 4–7. Alternatively, users may obtain such software from the principal author of this guide at a nominal cost.

References

Ayoub, M.M. and Mital, A., 1989. *Manual Materials Handling*. London: Taylor & Francis.

Jiang, B.C. and Mital, A., 1986. A procedure for designing/evaluating manual materials handling tasks. *International Journal of Production Research*, **24**(4), 913–925.

Mital, A., 1983. Generalized model structure for evaluating/designing manual materials handling jobs. *International Journal of Production Research*, **21**(3), 401–412.

Mital, A., 1991. Design and analysis of multiple activity manual materials handling tasks. In *Industrial Ergonomics: Case Studies* ed. by B. Mustafa and D.C. Alexander, pp. 29–40. Norcross: Industrial Engineering and Management Press.

Mital, A., 1995. Using "A Guide to Manual Materials Handling" for designing/evaluating multiple activity manual materials handling tasks. *Proceedings of the IEA World Conference on Ergonomic Design, Interfaces, Products, and Information*, pp. 550–553.

Mital, A. and Ramakrishnan, A., 1996. A complex manual materials handling activity: comparison of capacity data and outcomes of popular design guidelines. *Proceedings of the Fourth Pan-Pacific Conference on Ergonomics*, Taipei, Taiwan, November 1996.

Waters, T.R., Putz-Anderson, V., Garg, A., and Fine, L.J., 1993. Revised NIOSH equation for the design and evaluation of manual lifting tasks. *Ergonomics*, **36**, 749–776.

Chapter 11

High and very high frequency manual lifting/lowering and carrying (load transfer)

Introduction

Despite the fact that MMH activities have long been recognized as the primary source of severe and costly overexertion injuries in industry, and a variety of design guidelines have been proposed (see Ayoub and Mital, 1989; Mital *et al.*, 1993; and Chapter 1 of this guide), little attention has been paid to addressing high and very high frequency manual handling tasks (14 times per minute and higher). In fact, the highest frequency considered has generally been 12 times per minute, or lower.

The rapid world-wide growth in express air cargo and packaging services in the last 10–12 years, and in the trucking industry, has necessitated handling of packages at a very high rate. Moving 1200 or more packages per hour per person for up to 2 hours, and occasionally for 3 and 4 hours, is the norm for these industries. Since these packages vary widely in weight (as high as 32 kg) and size and may also involve a variety of movements (lifting, carrying, etc.; owing to such high frequencies, carrying activities are better described as load transferring activities), the work rates generated are very high.

Even though injury statistics breakdown for these kinds of tasks is presently unavailable, the high pace, and consequently the high work rate at which these tasks are performed, puts a very heavy cardiovascular and metabolic energy demand on the workers performing these tasks. The workers performing such tasks, thus, are exposed to considerable risk and could suffer from severe overexertion injuries (Mital and Founooni-Fard, 1994).

Since the tasks are performed at very high frequencies, psychophysical and physiological design approaches appear most appropriate for assessing capabilities of workers performing such tasks (see Chapter 3 for a description of various design criteria).

Weight limits for high and very high frequency manual handling tasks

The weight limits provided in this section are based on the studies by Founooni-Fard and Mital (1993a,b) and Mital and Founooni-Fard (1994).

Lifting and lowering

The manual lifting capability data (maximum acceptable weight of lift; MAWL) are given in Table 11.1. The heart rates and oxygen consumption values corresponding to Table 11.1 MAWL values are given in Table 11.2. The corresponding values of the ratings of perceived exertion (RPE) of the shoulder, back, and whole body for lifting are given in Table 11.3. The physiological responses and RPE values for lowering MAWL are provided in Tables 11.4 and 11.5, respectively.

Even though Founooni-Fard and Mital (1993a) did not include females in their study, on the basis of our discussion in Chapter 2

Table 11.1 Recommended weight of lift[a, b] (kg) for male industrial workers for high and very high frequencies (lifts/minute)

Population percentile	14/min	16/min	18/min	20/min	22/min
90	7.6	6.6	6.6	6.6	5.2
75	9.8	8.4	8.4	8.4	6.8
50	12.2	10.4	10.2	9.8	8.6
25	14.6	12.4	12.0	10.2	10.2
10	16.8	14.2	13.8	13.0	12.0

[a]These recommendations are for a working duration of 2 hours. Longer working durations are not recommended unless the work rate (by reducing either the frequency or the weight or both) is reduced.
[b]The values are for a box size of 30.5 cm. For other box sizes, use correction factors based on proportionate interpolation from Table 4.2 in Chapter 4.

Table 11.2 Heart rate (bpm; upper row) and oxygen uptake (VO_2 in litre/minute; lower row) corresponding to MAWL values given in Table 11.1 (lifting)

Population percentile	14/min	16/min	18/min	20/min	22/min
90	127	120	123	133	126
	1.41	1.45	1.41	1.45	1.45
75	139	135	136	145	143
	1.69	1.77	1.67	1.61	1.67
50	HR = 153	HR = 152	HR = 151	HR = 159	HR = 162
	VO_2 = 2.01	VO_2 = 2.13	VO_2 = 1.98	VO_2 = 1.90	VO_2 = 2.03
25	167	169	166	173	181
	2.33	2.49	2.30	2.20	2.40
10	179	178	179	183	198
	2.61	2.81	2.55	2.55	2.81

Table 11.3 Ratings of perceived exertion (RPE) of the shoulder (upper row), back (middle row), and whole body (lower row) corresponding to MAWL values given in Table 11.1 (lifting)

Population percentile	14/min	16/min	18/min	20/min	22/min
90	RPES = 10	8	9	9	9
	RPEB = 13	13	12	14	14
	RPEWB = 12	12	12	12	13
75	RPES = 10	9	10	10	10
	RPEB = 14	13	13	14	14
	RPEWB = 12	12	13	13	13
50	RPES = 11	10	11	11	11
	RPEB = 15	14	14	15	15
	RPEWB = 13	13	14	13	14
25	RPES = 11	11	12	12	12
	RPEB = 16	15	15	16	16
	RPEWB = 14	14	15	14	15
10	RPES = 12	12	13	13	13
	RPEB = 17	15	16	16	16
	RPEWB = 14	14	16	14	15

Table 11.4 Heart rate (bpm; upper row) and oxygen consumption (VO_2 in litre/minute; lower row) corresponding to MAWL values given in Table 11.1 (lowering)

Population percentile	14/min	16/min	18/min	20/min	22/min
90	105	106	115	117	111
	1.11	0.77	1.16	1.13	1.29
75	118	124	131	132	128
	1.37	1.16	1.45	1.51	1.66
50	HR = 133	HR = 144	HR = 149	HR = 148	HR = 148
	VO_2 = 1.65	VO_2 = 1.60	VO_2 = 1.78	VO_2 = 1.94	VO_2 = 2.08
25	148	164	167	164	168
	1.93	2.04	2.11	2.33	2.50
10	161	182	183	179	185
	2.19	2.43	2.40	2.75	2.87

Table 11.5 Ratings of perceived exertion (RPE) of the shoulder, back, and whole body corresponding to MAWL values given in Table 11.1 (lowering)

Population percentile	14/min	16/min	18/min	20/min	22/min
90	RPES = 9	8	9	9	9
	RPEB = 11	12	12	12	13
	RPEWB = 10	11	12	10	12
75	RPES = 10	9	10	10	10
	RPEB = 12	13	12	13	14
	RPEWB = 11	12	13	11	13
50	RPES = 10	10	11	11	11
	RPEB = 13	14	13	14	15
	RPEWB = 12	13	13	12	14
25	RPES = 10	11	12	12	12
	RPEB = 14	15	14	15	15
	RPEWB = 13	14	13	13	14
10	RPES = 11	12	12	13	13
	RPEB = 15	15	15	16	16
	RPEWB = 14	15	14	14	15

Table 11.6 Recommended weight of carry[a,b](kg) for male industrial workers for high and very high frequencies (carries/minute)

Population percentile	14/min	16/min	18/min	20/min	22/min
90	10.3	10.3	10.3	10.3	8.9
75	12.9	12.9	12.6	12.2	11.0
50	15.8	15.8	15.0	14.2	13.4
25	18.7	18.7	17.4	16.2	15.8
10	21.3	20.9	19.5	17.3	17.3

[a]These recommendations are for a working duration of 2 hours. Longer working durations are not recommended unless the work rate (by reducing either the frequency or the weight or both) is reduced.
[b]Values are for a 30.5 cm box. If the box size is doubled, the carrying capacity should be reduced by 4% (see Chapter 7 for discussion on load size in carrying activities).

Table 11.7 Heart rate (bpm; upper row) and oxygen uptake (VO_2 in litre/minute; lower row) corresponding to MAWL values given in Table 11.6

Population percentile	14/min	16/min	18/min	20/min	22/min
90	111	118	124	127	127
	0.74	1.0	0.96	1.07	1.1
75	122	127	133	137	137
	1.0	1.22	1.22	1.33	1.33
50	HR = 134	HR = 137	HR = 143	HR = 147	HR = 150
	VO_2 = 1.29	VO_2 = 1.42	VO_2 = 1.51	VO_2 = 1.62	VO_2 = 1.63
25	146	147	153	157	163
	1.58	1.62	1.80	1.91	1.93
10	157	156	152	167	173
	1.84	1.84	2.06	2.06	2.16

(subsection 'Gender') it is recommended that the maximum weights for females for corresponding frequencies never exceed 75% of the values given in Table 11.1 for males. *In fact, it is highly desirable that, owing to differences in the aerobic capacities of men and women (Astrand and Rodahl, 1986), the maximum weights of lift for females should not exceed 65% (or ⅔) of the values given in Table 11.1.*

Carrying and turning (load transfer)

Table 11.6 shows the manual carrying capability data for males. The corresponding values of heart rate and oxygen uptake are given in Table 11.7. The RPE values of the arms, shoulders, back, and whole body at the maximum acceptable weight of carry (MAWC) are given in Table 11.8.

Table 11.8 Ratings of perceived exertion (RPE) of the arms (upper row), shoulder (second row), back (third row), and whole body (lower row) corresponding to MAWL values given in Table 11.6

Population percentile	14/min	16/min	18/min	20/min	22/min
90	RPEA = 9	10	11	11	12
	RPES = 8	9	9	10	10
	RPEB = 8	11	11	11	11
	RPEWB = 9	11	11	11	11
75	RPEA = 11	11	12	12	12
	RPES = 9	10	10	10	11
	RPEB = 9	11	11	11	12
	RPEWB = 10	11	11	12	12
50	RPEA = 12	12	12	12	13
	RPES = 10	11	11	11	12
	RPEB = 11	12	12	12	13
	RPEWB = 11	12	12	13	13
25	RPEA = 13	13	13	13	14
	RPES = 11	12	12	12	13
	RPEB = 13	13	13	13	14
	RPEWB = 12	13	12	14	14
10	RPEA = 14	14	13	14	15
	RPES = 12	12	13	13	14
	RPEB = 14	13	13	14	15
	RPEWB = 13	14	13	15	15

As in the case of lifting and lowering, the values given in Table 11.6 must be reduced for females. The recommended weight of carry for females must not exceed 75% of the values given in Table 11.6; preferably, the values for females should not exceed 65% of the values for males (see Chapter 2 for discussion).

Table 11.9 Endurance time[a] (minutes) for males as a function of lifting frequency (lifts/minute), load, and lifting height

Load (kg)	Lifting height	Population percentile	12/min	14/min	16/min
5	Floor–80 cm	90	295	239	183
		75	352	308	265
		50	415	385	356
		25	478	462	447
		10	480	480	480
	80–132 cm	90	374	358	343
		75	411	398	386
		50	453	443	433
		25	480	480	480
		10	480	480	480
	132–183 cm	90	260	213	166
		75	320	282	244
		50	388	360	333
		25	456	439	422
		10	480	480	480
10	Floor–80 cm	90	142	116	91
		75	232	120[b]	120[b]
		50	333	226[b]	120[b]
		25	434	277[b]	120[b]
		10	480	314[b]	149[b]
	80–132 cm	90	272	248	224
		75	332	311	290
		50	398	381	364
		25	464	451	438
		10	480	480	480
	132–183 cm	90	91	61	32
		75	188	155	122
		50	296	263	230
		25	404	371	338
		10	480	454	428
15	Floor–80 cm	90	29	14	0
		75	129	64[b]	0[b]
		50	155[b]	120[b]	0[b]
		25	180[b]	120[b]	60[b]
		10	281[b]	200[b]	120[b]
	80–132 cm	90	110	68	26
		75	192	143	95
		50	284	228	172
		25	376	312	249
		10	458	388	318
	132–183 cm	90	0	1	2
		75	70	50	31
		50	148	106	64
		25	226	161	97
		10	296	211	126

[a]Maximum endurance time limited to 8 hours (480 minutes) working time.
[b]Endurance time revised for recommended weight limit data in Table 11.1.

Endurance time at high lifting frequencies

Endurance time (ET) may simply be defined as the duration during which the performance of a task can be sustained. There are occasions when the loads are relatively light (compared with recommended weights in Table 11.1) and it is of interest to know how long a person can perform the manual lifting task at a specified frequency. Asfour *et al.* (1991) developed ET tables for different frequencies and lifting height ranges. Table 11.9 provides endurance time data (in minutes) for males for different load, frequency, and height range combinations. Linear interpolation and extrapolation may be used to obtain ET values for neighbouring frequencies (up to 18 lifts/minute) and loads. (The loads and durations must conform to Table 11.1 recommendations. Also, note that the upper limit on load for females is 20 kg; see Chapters 3 and 4.) The study does not provide ET data for females. **Given the difference in the aerobic capacity of males and females (Astrand and Rodahl, 1986), it is recommended that endurance time for females must not exceed 75% of the ET for males; preferably, the female endurance time should be limited to 65% of the male endurance time (refer to discussion in Chapter 2).**

References

Asfour, S.S., Khalil, T.M., Genaidy, A.M., Akein, M., Jomoah, I.M., Koshy, J.G., and Tritar, M., 1991. *Ergonomic Injury Control In High Frequency Lifting Tasks.* Final Report, Grant No. 5-R01-OH02591-02, Cincinnati: National Institute for Occupational Safety and Health.

Astrand, P.-O. and Rodahl, K., 1986. *Textbook of Work Physiology: Physiological Basis of Exercise.* New York: McGraw-Hill.

Ayoub, M.M. and Mital, A., 1989. *Manual Materials Handling.* London: Taylor & Francis.

Founooni-Fard, H. and Mital, A., 1993a. A psychophysiological study of high and very high frequency manual materials handling: Part I – lifting and lowering. *International Journal of Industrial Ergonomics*, **12**, 127–141.

Founooni-Fard, H. and Mital, A., 1993b. A psychophysiological study of high and very high frequency manual materials handling: Part III – Carrying and turning. *International Journal of Industrial Ergonomics*, **12**, 143–152.

Mital, A., Founooni-Fard, H. and Brown, M.L. 1994. Physical fatigue in high and very high frequency manual materials handling: perceived exertion and physiological indicators. *Human Factors*, **36**, 219–231.

Mital, A., Nicholson, A.S., and Ayoub, M.M., 1993. *A Guide to Manual Materials Handling.* London: Taylor & Francis.

Chapter 12

Determination of rest allowances

Introduction

Performance of tasks depends upon muscular contractions, which in turn require expenditure of metabolic energy. The metabolic energy requirements are substantial in the event the task being performed is physical and highly repetitive. With continuous work, the muscular contractions become weak as the metabolic energy resources are depleted. Muscles also become tired and ache when significant amounts of lactic acid are accumulated within them. Whenever these conditions occur, a state of fatigue, known as physiological muscle fatigue, results and the level and quality of work output starts declining (Mital and Founooni-Fard, 1994). Further, as the muscular system weakens, the individual becomes increasingly susceptible to overexertion injury. Rest allowances are given to avoid these detrimental effects. During rest periods, workers recover from the fatigue that is generated as a result of producing work. (For further discussion on fatigue/rest allowances, refer to Mital *et al.* (1991) and Ayoub and Mital (1989; Chapter 9).)

Types of allowances

There are three basic types of allowances that are provided to workers:

- personal allowances – for needs such as getting a drink of water, going to toilet, etc.
- delay allowances – for work-related interruptions such as talking to the supervisor, equipment breakdowns, etc.
- Fatigue allowances – for recovery from work-related demands.

As stated above, fatigue or rest allowances are given so that workers may recover from physiological and psychological effects of work. Rest allowances can be added to the working time either as a percentage of work time or as a percentage of shift time. This chapter discusses a procedure that provides rest allowances as a percentage of shift time. (For details about the concept, the model, and its validation, refer to Ayoub and Mital (1989).)

Rest allowance determination procedure

The procedure utilizes the total daily energy requirement concept. The total energy requirement is adjusted for age and energy required for food ingestion. The calculation of basal and leisure metabolism is based on the hours during which these activities are performed (default is 8 hours). The work (shift) duration is also taken into consideration. It is assumed that there is a limit to the amount of metabolic energy available to produce work. The entire rest allowance determination procedure is shown in Figure 12.1.

In order to use this procedure, the following inputs are required:

1. worker gender (M or F)
2. worker age (years)
3. worker body-weight (pounds)
4. number of hours of sleep per day (hours)
5. shift duration (hours)

6. number of tasks performed during the shift
7. time duration of each task (hours)
8. metabolic energy requirement for each task (kcal/hour)
9. an estimate of worker's physiological condition (high, average, or low).

To increase the accuracy of the procedure two additional pieces of information are needed:

10. worker aerobic capacity (ml/min/kg)
11. average energy requirement/min when not working or sleeping (kcal).

In the absence of these, the determined rest allowances will be conservative.

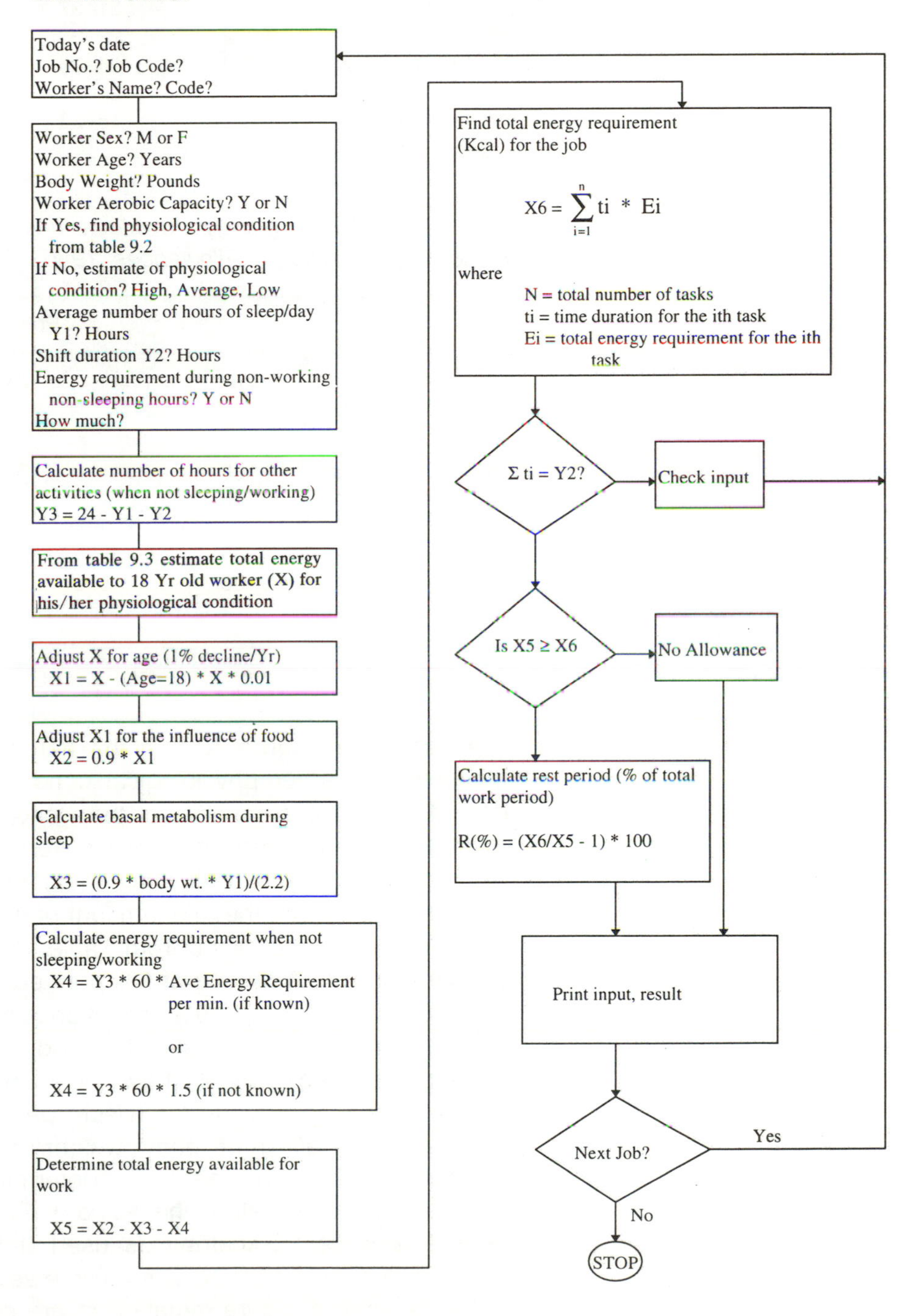

Figure 12.1 Rest allowance determination procedure (Case study 1 – Mital and Shell, 1984).

Table 12.1 The range of expected maximal aerobic capacity, by sex and age (American Heart Association, 1972)

Age (years)	Maximal oxygen uptake (ml/kg/min)				
	Low	Fair	Average	Good	High
			Men		
20–29	25	25–33	34–42	43–52	53+
30–39	23	23–30	31–38	39–49	49+
40–49	20	20–26	27–35	36–44	45+
50–59	18	18–24	25–33	34–42	43+
60–69	16	16–22	23–30	31–40	41+
			Women		
20–29	24	24–30	31–37	38–48	48+
30–39	20	20–27	28–33	34–44	45+
40–49	17	17–23	24–30	31–41	42+
50–59	15	15–20	21–27	28–37	38+
60–69	13	13–17	18–23	24–34	35+

Table 12.2 Total energy available/day (kcal) to an 18-year-old (Recommended Dietary Allowances, 1980)

		Physical condition				
Sex	Aerobic capacity	Low	Fair	Average	Good	High
Male	Known	2100	2575	3050	3225	4000
	Not known	2100	–	3050	–	4000
Female	Known	2100	2325	2550	2775	3000
	Not known	2100	–	2550	–	3000

As shown in Figure 12.1, if the aerobic capacity of the worker is known then his/her physiological condition is calculated from Table 12.1. If the aerobic capacity of the worker is not known, a qualitative estimate (low, average, or high; smokers, for instance, may have low aerobic capacity) is needed. Once the general physiological condition of the worker is known, the total amount of energy available to an 18-year old, per day, is determined from Table 12.2. Next, this total amount is adjusted for age. A decline of 1% per year, above 18 years, is used as the adjustment factor. A 10% reduction is made in this value to allow for the energy required in digesting the food.

Once these adjustments have been made, basal metabolism and energy expenditure during the leisure period are calculated by using a value of 0.9 kcal/kg of body weight/hour of sleep (corrected for metabolic savings in sleep). To determine the energy requirement during leisure time, either the actual value (in kcal/min) is used or a default value of 1.5 kcal/min is used (this value represents energy expenditure during a typical leisure time sedentary task).

The basal and leisure metabolism are subtracted from the adjusted total energy to determine the amount available for producing work. Next, the total energy requirement for the job is determined as follows:

$$\text{Total energy requirement for all jobs (kcal)} = \sum_{i=1}^{N} t_i E_i$$

where t_i is time (in hours) for the ith job, E_i is its energy requirement in kcal, and N is total number of jobs carried out during the shift. Next, the rest period, as a percent of the shift duration, is determined as follows:

Rest (%) = [(total energy requirement for the job/adjusted, total energy available for work) − 1] × 100

The procedure described here is applicable to any shift duration and any combination of physical tasks.

In order to simplify the usage, an interactive computer program (written in Basic) has been developed. A listing of this program is given at the end of this chapter.

Table 12.3 Program output for the above example

Date:	7/3/84
Job number:	Example
Job code:	Case 1
Worker name:	Mr. X
Sex:	M
Age (years):	29
Weight in pounds:	145
Aerobic capacity in ml/kg/min:	55
Physiological condition computed from the table:	5
Hours of sleep per day:	8
Shift duration (hours):	8
Energy available for 18-year old from Table 12.2, in kcal:	4000
Energy required when not sleeping and not working, in kcal:	816
Total energy available for work (kcal):	1913.45
Total number of jobs:	2
Time duration of job (hours):	Energy required for job(kcal/h)
7	237.6
1	102
Total energy requirement of the job (kcal):	1765.2
Rest period as a percent of shift duration:	0

Program output

Date:	7/3/84
Job number:	Example
Job code:	Case 2
Worker name:	Mr. X
Sex:	M
Age (years):	29
Weight in pounds:	145
Aerobic capacity in ml/kg/min:	55
Physiological condition computed from the table:	5
Hours of sleep per day:	8
Shift duration (hours):	12
Energy available for 18-year old from Table 12.2, in kcal:	4000
Energy required when not sleeping and not working, in kcal:	408
Total energy available for work (kcal):	2321.45
Total number of jobs:	2
Time duration of job (hours):	Energy required for job (kcal/h)
7	206
1	102
Total energy requirement for the job (kcal):	2290
Rest period as a percent of shift duration:	0

Table 12.4 Program output with known and unknown aerobic capacity (Ayoub and Mital, 1989)

Aerobic capacity known

Date:	12/3/83
Job number:	1
Job code:	Example
Worker name:	Mr. X
Sex:	M
Age (years):	35
Weight in pounds:	150
Aerobic capacity in ml/kg/min:	44
Physiological condition computed from the table:	4
Hours of sleep per day:	7
Shift duration (hours):	8
Energy available for 18-year old from Table 12.2, in kcal:	3525
Energy required when not sleeping and not working, in kcal:	810
Total energy available for work (kcal):	1393.63
Total number of jobs:	3
Time duration of job (hours):	Energy required for job (kcal/h)
2	180
3	200
3	200
Total energy requirement of the job (kcal):	1560
Rest period as a percent of shift duration:	11.9397

Aerobic capacity unknown

Date:	12/3/83
Job number:	1
Job code:	Example
Worker name:	Mr. X
Sex:	M
Age (years):	35
Weight in pounds:	150
Estimated physiological condition:	3
Hours of sleep per day:	7
Shift duration (hours):	8
Energy available for 18-year old from Table 12.2, in kcal:	3050
Energy required when not sleeping and not working, in kcal:	810
Total energy available for work (kcal):	1038.8
Total number of jobs:	3
Time duration of job (hours):	Energy required for job (kcal/h)
2	180
3	200
3	200
Total energy requirement for the job (kcal):	1560
Rest period as a percent of shift duration:	50.1726

Examples

This section shows program output for an actual case as well as the differences in rest allowance determination when the aerobic capacity is known and when it is not known.

The individual involved is a male with the following particulars:

Age = 29 years
Body weight = 145 lbs
Aerobic capacity = 55 ml/kg/min
Hours of sleep/day = 8
Shift duration (hours) = 8 (case 1) and 12 (case 2)
Energy requirement during leisure time (kcal/min) = 1.7
Energy required for the job (kcal/h) = 237.6 (= 60 min x 3.96 kcal/min) for case 1, = 206 (= 60 min x 3.44 kcal/min) for case 2.

Two different cases are examined: case 1, 8 h shift; case 2, 12 h shift. Table 12.3 shows the program output for the two cases. In neither case is a rest allowance needed. The total energy requirement for the job is quite close to the total energy available for producing work.

Table 12.4 shows program output when the aerobic capacity is known and when it is not known and is estimated from physiological condition.

A comparison of rest allowances when aerobic capacity is known and when it is estimated shows that rest allowances determined in the absence of accurate knowledge of aerobic capacity are quite conservative. Therefore, whenever possible, the aerobic capacity value should be entered. The aerobic capacity determination techniques described in the manual prepared by Mital (1996) for NIOSH may be used for this purpose. The manual can be obtained from NIOSH at no cost.

References

American Heart Association, 1972. *Exercise Testing and Training of Apparently Healthy Individuals: A Handbook for Physicians.* American Heart Association.

Ayoub, M.M. and Mital, A., 1989. *Manual Materials Handling.* London: Taylor & Francis.

Mital, A., 1996. *Recognition of Musculoskeletal Injury Hazards.* Contract No. CDC-94071VID. Cincinnati: National Institute of Occupational Safety and Health.

Mital, A. and Founooni-Fard, H., 1994. Physical fatigue in high and very high frequency manual materials handling: perceived exertion and physiological indicators. *Human Factors*, **36**, 219–231.

Mital, A., Bishu, R.R., and Manjunath, S.G., 1991. Review and evaluation of techniques for determining fatigue allowances. *International Journal of Industrial Ergonomics*, **8**, 165–178.

Mital, A. and Shell, R.L., 1984. A comprehensive metabolic energy model for determining rest allowances for physical tasks. *Journal of Methods Time Measurement*, **XI**, 2–8.

Recommended Dietary Allowances, 1980. Washington, DC: Food and Nutrition Board, National Academy of Sciences, National Research Council.

Rest allowance determination program source listing

```
10 '    PROGRAM TO COMPUTE THE REST PERIOD AS A PERCENT OF TOTAL WORK PERIOD BY
20 '    ANIL MITAL
30 '    UNIVERSITY OF CINCINNATI, OHIO
40 '
50 '
60 '    THIS PROGRAM COMPUTES THE REST PERIOD AS A PERCENT OF TOTAL WORK PERIOD
70 '    TWO LOOKUP TABLES ARE IMPLEMENTED IN THE PROGRAM GIVEN THE AGE AND
80 '    AEROBIC CAPACITY OF THE WORKER, IT COMPUTES THE PHYSIOLOGICAL CONDITION
90 '    WHICH COULD FALL IN FIVE CATEGORIES.
100'
110'    SECOND TABLE COMPUTES TOTAL AEROBIC ENERGY FOR 18 YEAR OLD MALE OR FEMALE
120'    FOR 24 HOURS WHEN THE PHYSIOLOGICAL CONDITION IS KNOWN
130'
140'
150     DIM T(20), E(20), MX(5), FX(5)
160     DIM WCAP1(5), WCAP2(5), WCAP3(5), WCAP4(5), WCAP5(5)
170     DIM MCAP1(5), MCAP2(5), MCAP3(5), MCAP4(5), MCAP5(5)
180'
190'
200'    THE FOLLOWING IS THE TABLE FOR 18 YEAR OLD MALE (MX) AND FEMALE (FX)
210     MX(1)=2100: MX(2)=2575: MX(3)=3050: MX(4)=3525: MX(5)=4000
220     FX(1)=2100: FX(2)=2325: FX(3)=2550: FX(4)=2775: FX(5)=3000
230'
240'
250'    TABULAR VALUES FOR WOMEN'S PHYSIOLOGICAL CONDITION DEPENDING UPON AGE
260'    AND AEROBIC CAPACITY
270'
280     WCAP1(1)=24: WCAP1(2)=30: WCAP1(3)=37: WCAP1(4)=48: WCAP1(5)=100
290     WCAP2(1)=20: WCAP2(2)=27: WCAP2(3)=33: WCAP2(4)=44: WCAP2(5)=100
300     WCAP3(1)=17: WCAP3(2)=23: WCAP3(3)=30: WCAP3(4)=41: WCAP3(5)=100
310     WCAP4(1)=15: WCAP4(2)=20: WCAP4(3)=27: WCAP4(4)=37: WCAP4(5)=100
320     WCAP5(1)=13: WCAP5(2)=17: WCAP5(3)=23: WCAP5(4)=34: WCAP5(5)=100
330'
340'
350'    TABULAR VALUES FOR MEN'S PHYSIOLOGICAL CONDITION DEPENDING UPON AGE
360'    AND AEROBIC CAPACITY
370'
380     MCAP1(1)=25: MCAP1(2)=33: MCAP1(3)=42: MCAP1(4)=52: MCAP1(5)=100
390     MCAP2(1)=23: MCAP2(2)=30: MCAP2(3)=33: MCAP2(4)=49: MCAP2(5)=100
400     MCAP3(1)=20: MCAP3(2)=26: MCAP3(3)=35: MCAP3(4)=44: MCAP3(5)=100
410     MCAP4(1)=18: MCAP4(2)=24: MCAP4(3)=33: MCAP4(4)=42: MCAP4(5)=100
420     MCAP5(1)=16: MCAP5(2)=22: MCAP5(3)=30: MCAP5(4)=40: MCAP5(5)=100
430'
440'
450     AGE1=29: AGE2=39: AGE3=49: AGE4=59: AGE5=69
460'
470'
480     INPUT "DATE : ", DATE$
490     INPUT "JOB NUMBER: ", JOB$
500'
510     INPUT "JOB CODE", JOB CODE$
520     INPUT "WORKERNAME : ", W. NAME$
530     INPUT "SEX (M-MALE, F-FEMALE) : ", SEX$
540     IF SEX$="M" OR SEX$="F" THEN GOTO 570
550     GOTO 530
560'
570     INPUT "AGE (BETWEEN 20 AND 69): ", AGE
580     IF AGE <19 OR AGE>69 THEN GOTO 570
590'
600     INPUT "WEIGHT IN POUNDS: ", WEIGHT
610     IF WEIGHT <1 THEN GOTO 600
```

```
620'
630'
640        INPUT "IS AEROBIC CAPACITY KNOWN? (Y/N): ", AC
650        IF AC YESNO$="N" THEN GO TO 1260
660        INPUT "ENTER AEROBIC CAPACITY IN ML/KG/MIN (1-100):", ACAPACITY
670'
680'
690'       SINCE AEROBIC CAPACITY IS KNOWN, PHYSIOLOGICAL CONDITION IS COMPUTED
700'       FROM THE TABLE
710        IF SEX$= "M" GOTO 1000
720'
730'
740'       THIS SECTION COMPUTES PHYSIOLOGICAL CONDITION OF WOMEN
750'
760        IF AGE > AGE1 THEN GOTO 800
770        FOR I=1 TO 5
780        IF ACAPACITY < WCAP1(I) THEN GOTO 1210
790        NEXT I
800        IF AGE > AGE2 THEN GOTO 840
810        FOR I=1 TO 5
820        IF ACAPACITY < WCAP2(I) THEN GOTO 1210
830        NEXT I
840        IF AGE > AGE3 THEN GOTO 880
850        FOR I=1 TO 5
860        IF ACAPACITY < WCAP3(I) THEN GOTO 1210
870        NEXT I
880        IF AGE > AGE4 THEN GOTO 920
890        FOR I=1 TO 5
900        IF ACAPACITY < WCAP4(I) THEN GOTO 1210
910        NEXT I
920        FOR I=1 TO 5
930        IF ACAPACITY < WCAP5(I) THEN GOTO 1210
940        NEXT I
950'
960'
970'
980'       THIS SECTION COMPUTES PHYSIOLOGICAL CONDITION OF MEN
990'
1000       IF AGE > AGE1 THEN GOTO 1040
1010       FOR I=1 TO 5
1020       IF ACAPACITY < MCAP1(I) THEN GOTO 1210
1030       NEXT I
1040       IF AGE > AGE2 THEN GOTO 1080
1050       FOR I=1 TO 5
1060       IF ACAPACITY < MCAP2(I) THEN GOTO 1210
1070       NEXT I
1080       IF AGE > AGE3 THEN GOTO 1120
1090       FOR I=1 TO 5
1100       IF ACAPACITY < MCAP3(I) THEN GOTO 1210
1110       NEXT I
1120       IF AGE > AGE4 THEN GOTO 1160
1130       FOR I=1 TO 5
1140       IF ACAPACITY < MCAP4(I) THEN GOTO 1210
1150       NEXT I
1160       FOR I=1 TO 5
1170       IF ACAPACITY < MCAP5(I) THEN GOTO 1210
1180       NEXT I
1190'
1200'
1210       PHYS. COND=I
1220       PRINT "COMPUTED PHYSIOLOGICAL CONDITION: ", PHYS. COND.
1230       GOTO 1310
1240'
1250'
1260       PRINT "SINCE AEROBIC CAPACITY IS NOT KNOWN, ENTER ESTIMATED"
```

```
1270    INPUT "PHYSIOLOGICAL CONDITION (1=LOW, 2=AVERAGE,  3=HIGH)", PHYS. COND
1280    IF PHYS. COND<1 OR PHYS. COND>3 THEN GOTO 1270
1290'
1300'
1310    INPUT "HOURS OF SLEEP PER DAY (Y1): ", Y1
1320    INPUT "SHIFT DURATION IN HOURS (Y2): ", Y2
1330    LET Y3=24-Y1-Y2
1340'
1350'
1360'   ENERGY AVAILABLE FOR 18 YEAR OLD IS COMPUTED FROM THE TABLE
1370    IF SEX$="F" THEN GOTO 1400
1380    X=MX (PHYS. COND)
1390    GOTO 1410
1400    X=FX (PHYS. COND)
1410    PRINT "ENERGY AVAILABLE TO 18 YEAR OLD: ", X
1420'
1430'
1440'   ADJUSTING FOR AGE - 1% DECLINE
1450    LET X1=X-(AGE-18)xXx.01
1460'
1470'
1480'   ADJUST X1 FOR THE INFLUENCE OF FOOD
1490    LET X2=.9xX1
1500'
1510'
1520'   CALCULATE BASAL METABOLISM DURING SLEEP
1530    LET X3=(.9xWEIGHTxY1)/2.2
1540'
1550'
1560'   CALCULATE X4, THE ENERGY REQUIREMENT WHEN NOT SLEEPING OR WHEN NOT
1570'   AT WORK
1580    INPUT "IS ENERGY (KCAL/MIN) WHEN NOT SLEEPING/WORKING KNOWN?", ER. YESNO$
1590    IF ER. YESNO$="Y" THEN INPUT "ENERGY REQUIRED: ", ER
1600    IF ER. YESNO$< >"Y" THEN GOTO 1630
1610    LET X4=Y3x60xER
1620    GOTO 1670
1630    LET X4=Y3x60x1.5
1640'
1650'
1660'   DETERMINE TOTAL ENERGY AVAILABLE FOR WORK (X5)
1670    LET X5=X2-X3-X4
1680    PRINT "X5=", X5
1690'
1700'
1710    INPUT "NUMBER OF JOBS N: ", N
1720    LET X6=0
1730    TT=0
1740    FOR I=1 TO N
1750    INPUT "ENTER TIME DURATION IN HOURS FOR THE ITH TASK:  ", T(I)
1760    INPUT "ENTER TOTAL ENERGY REQUIRED/HOUR FOR THE ITH TASK: ", E(I)
1770    LET X6=X6+T(I)xE(I)
1780    LET TT=TT+T(I)
1790    NEXT I
1800'
1810'
1820    IF TT=Y2 THEN GOTO 1860
1830    PRINT "TOTAL TIME DURATION (HOURS) IS NOT EQUAL TO  Y2 (SHIFT DURATION) CHECK
        INPUT"
1840    GOTO 640
1850'
1860    IF X5>X6 THEN R=0 ELSE R=(X6/X5-1)x100
1870    PRINT "REST PERIOD AS A PERCENT OF WORK PERIOD: ", R
1880'
1890    INPUT "DO YOU WANT A HARD COPY? (Y/N): ", HARD.YESNO$
1900    IF HARD YESNO$ < > "Y" GOTO 2280
```

```
1910'
1920    LPRINT
1930    LPRINT "PROGRAM OUTPUT"
1940    LPRINT
1941    LPRINT
1950    LPRINT "DATE: ", DATE$
1951    LPRINT
1960    LPRINT "JOB NUMBER: ", JOB$
1961    LPRINT
1970    LPRINT "JOB CODE: ", JOB.CODE$
1971    LPRINT
1980    LPRINT "WORKER NAME: ", W.NAME$
1981    LPRINT
1990    LPRINT "SEX: ", SEX$
1991    LPRINT
2000    LPRINT "AGE (YEARS): ", AGE
2001    LPRINT
2010    LPRINT "WEIGHT IN POUNDS: ", WEIGHT
2011    LPRINT
2020'
2030    IF AC. YESNO$="N" GOTO 2070
2040    LPRINT "AEROBIC CAPACITY INML/KG/MIN: ", ACAPACITY
2050    LPRINT "PHYSIOLOGICAL CONDITION COMPUTED FROM THE TABLE: ", PHYS. COND
2060    GOTO 2080
2070    LPRINT "ESTIMATED PHYSIOLOGICAL CONDITION: ", PHYS. COND
2080'
2081    LPRINT
2090    LPRINT "HOURS OF SLEEP PER DAY: ", Y1
2100    LPRINT "SHIFT DURATION (HOURS): ", Y2
2110'
2111    LPRINT
2120    PRINT "ENERGY AVAILABLE FOR 18 YEAR OLD FROM THE TABLE (X), IN KCAL: ", X
2130    LPRINT "ENERGY REQUIRED WHEN NOT SLEEPING AND NOT WORKING, IN KCAL: ", X4
2140    LPRINT "TOTAL ENERGY (KCAL) AVAILABLE FOR WORK: ", X5
2150    LPRINT
2151    LPRINT
2160    LPRINT "TOTAL NUMBER OF JOBS: ", N
2170    LPRINT
2180    LPRINT "TIME DURATION (HOURS) OF JOB                    ENERGY    REQUIRED
        (KCAL/HOUR) FOR JOB"
2190    FOR I=1 TO N
2200    LPRINT T(I), "                    ", E(I)
2210    NEXT I
2220'
2221    LPRINT
2230    LPRINT
2240    LPRINT "TOTAL ENERGY REQUIRED FOR THE JOB (KCAL): ", X6
2250    LPRINT "REST PERIOD AS A PERCENT OF SHIFT DURATION: " R
2260'
2270'
2280    INPUT "DO YOU WANT TO RUN THE NEXT JOB? (Y/N)", JOB.YESNO$
2290    IF JOB.YESNO$="Y" GOTO 480
2300    END
```

Chapter 13

Mechanical aids

It is not uncommon to encounter MMH tasks that are beyond the capability of one person and are performed so infrequently that redesigning them is not economically beneficial. Occasional handling of awkward and bulky objects is a task that would fit in this category. Under such circumstances, it is recommended that handling of the object be aided by some kind of mechanical equipment. Figures 13.3–13.28 show the widely available mechanical aids, not in any particular order, that may be used to assist workers in carrying out infrequent and physically demanding MMH activities that are not suitable for redesigning either for economical or technical reasons.

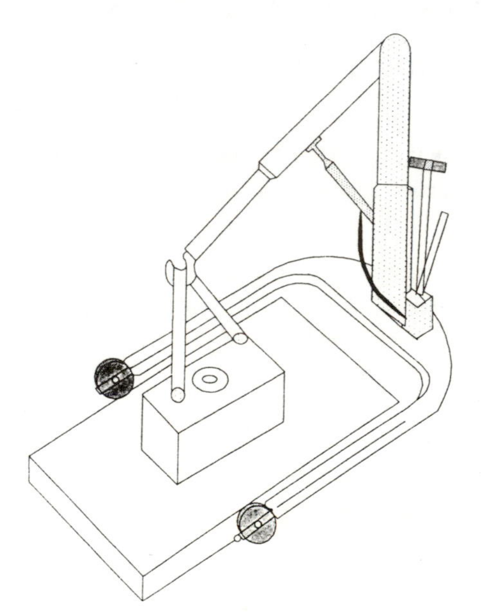

Figure 13.1 Portable crane.

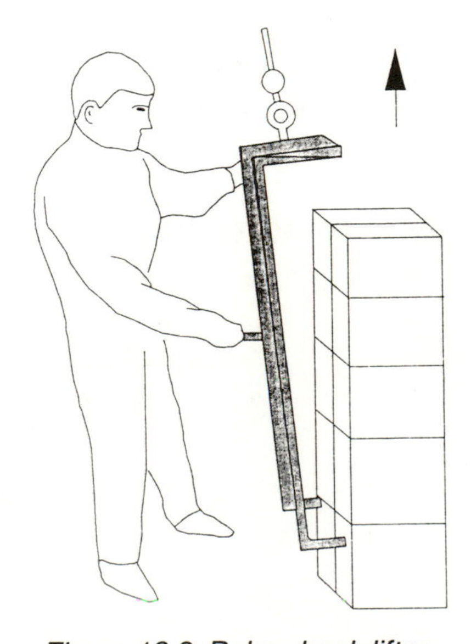

Figure 13.2 Below-hook lifter.

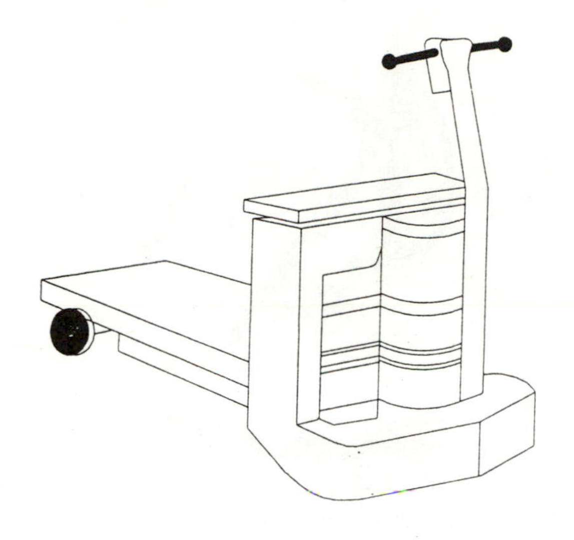

Figure 13.3 Platform truck.

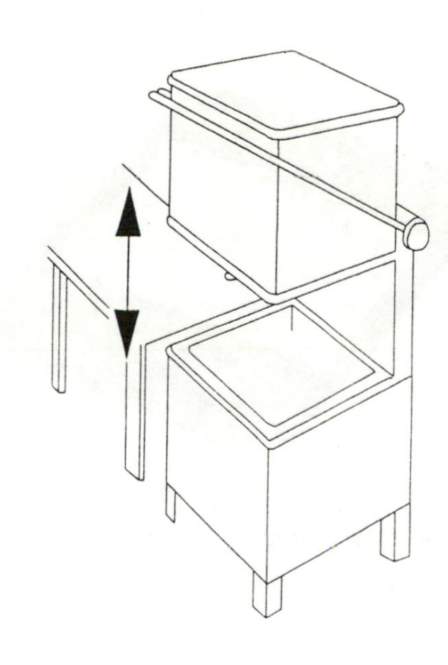

Figure 13.6 Box hauling device.

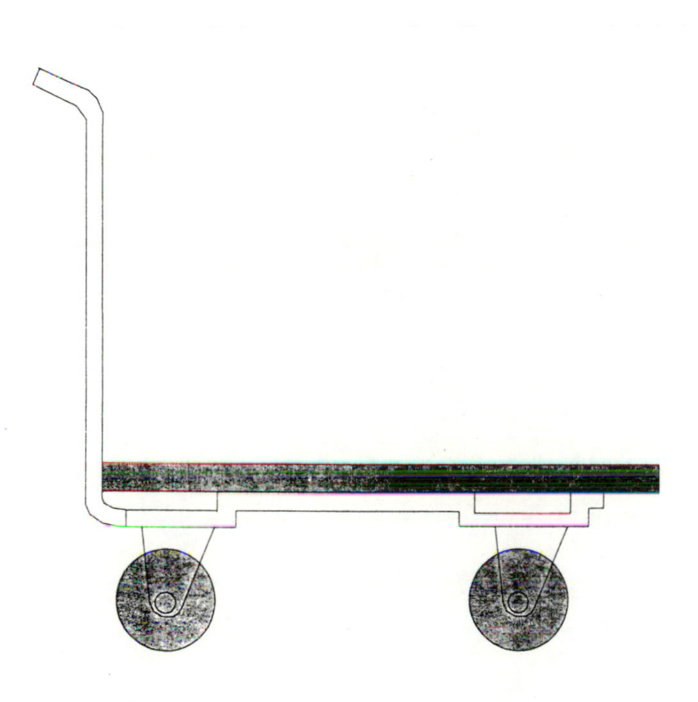

Figure 13.4 Four-wheel hand truck.

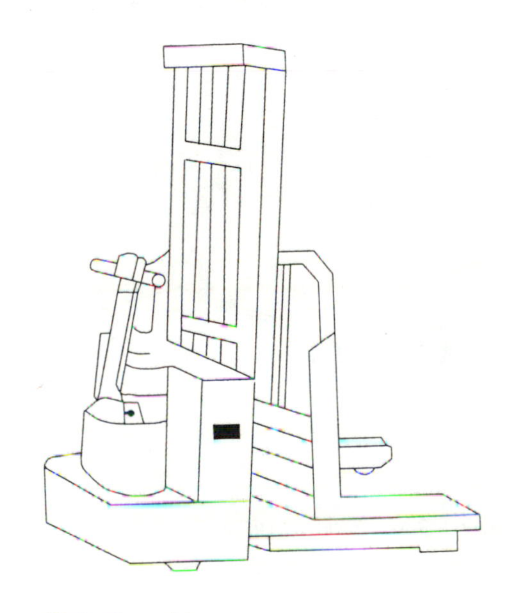

Figure 13.7 Portable elevator.

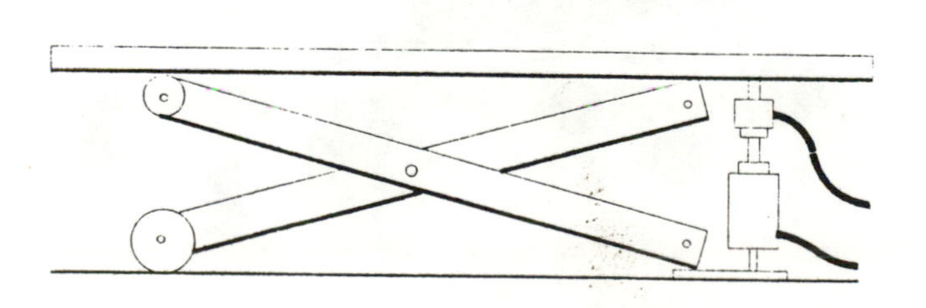

Figure 13.5 Lift table.

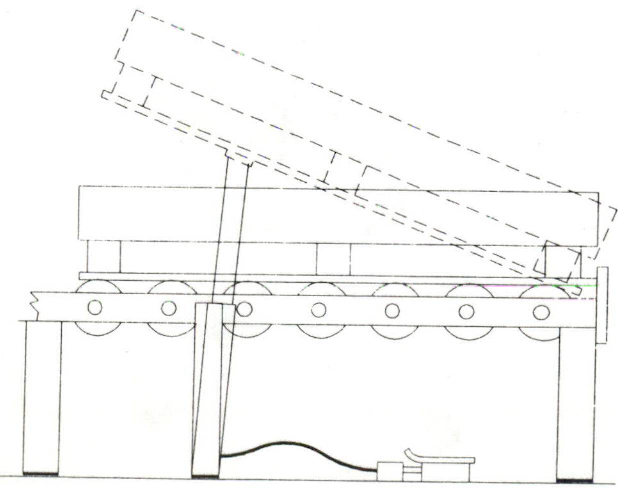

Figure 13.8 Conveyor and combined hydraulic dumping table.

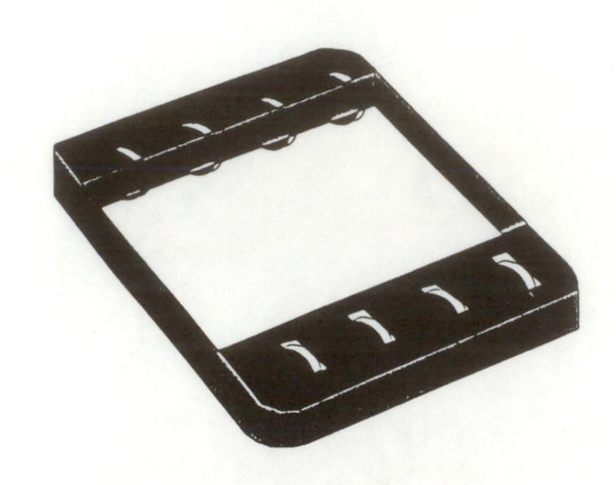

Figure 13.9 Pallet dolly.

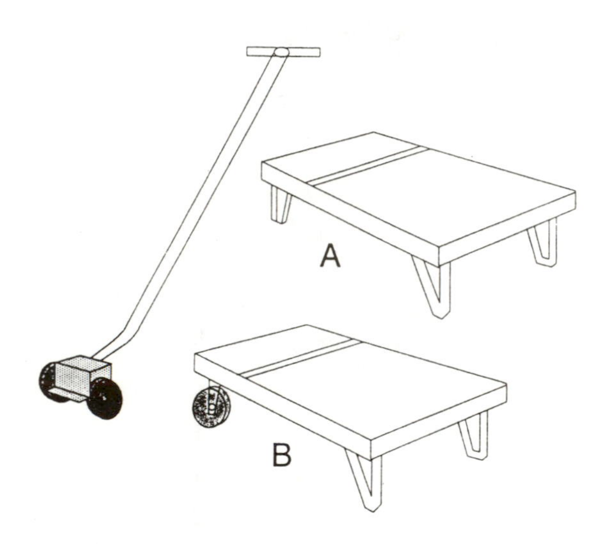

Figure 13.10 Skid truck.

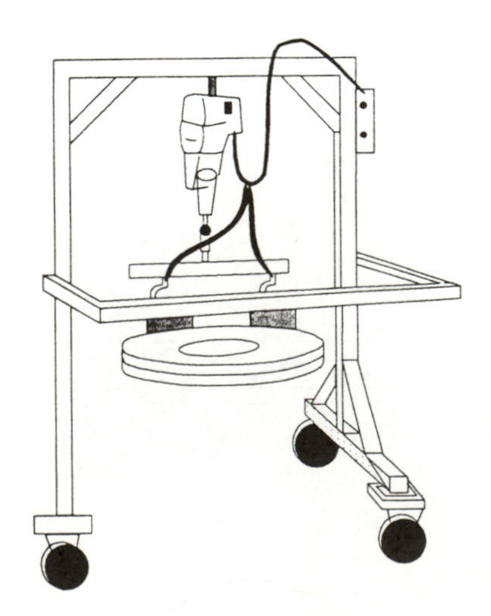

Figure 13.11 Lifting-beam type below-hook lifter.

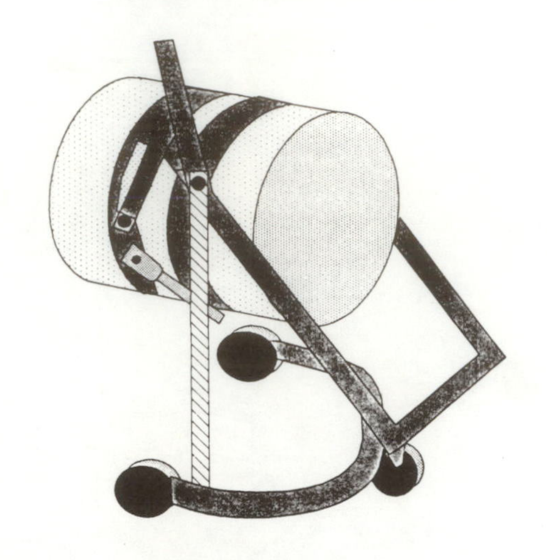

Figure 13.12 Tilting barrel carrier.

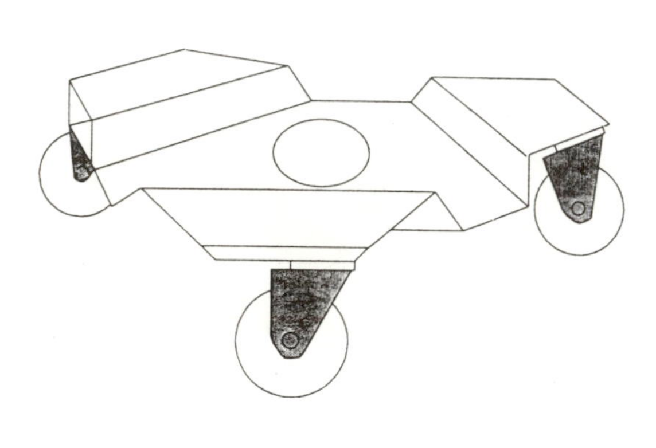

Figure 13.13 Tripod dolly.

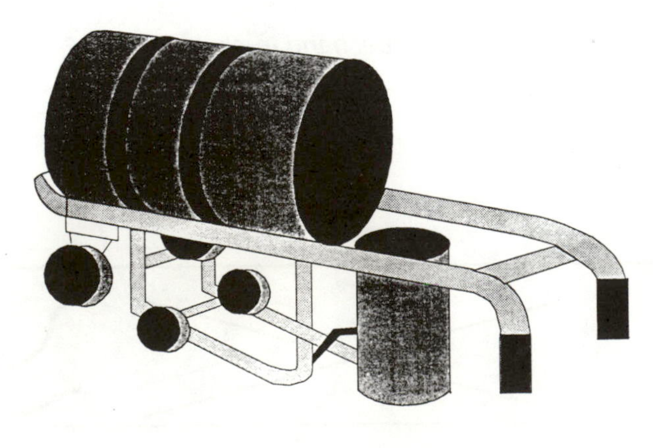

Figure 13.14 Drum truck and drainer.

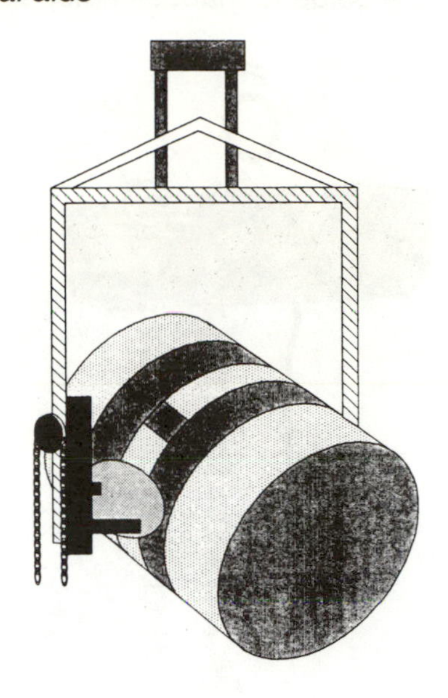

Figure 13.15 Drum lifter.

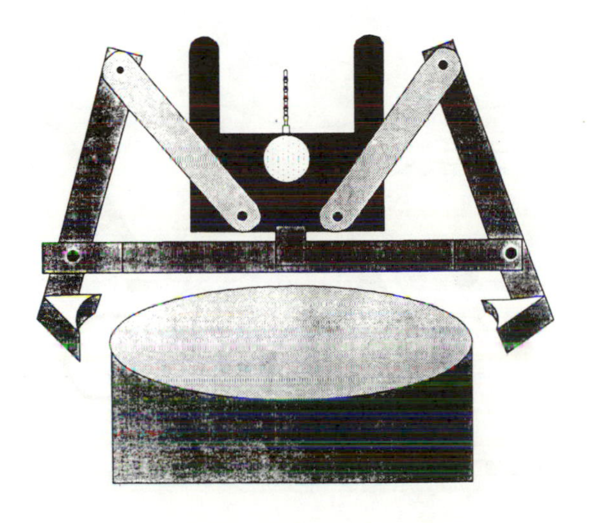

Figure 13.16 Vertical drum lifter.

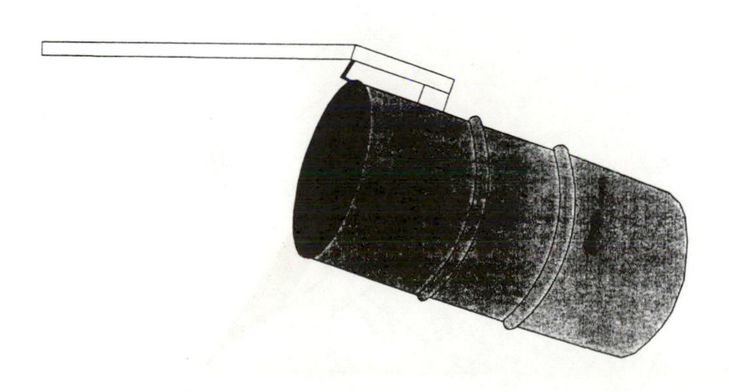

Figure 13.17 Drum upender.

Figure 13.18 Utility truck.

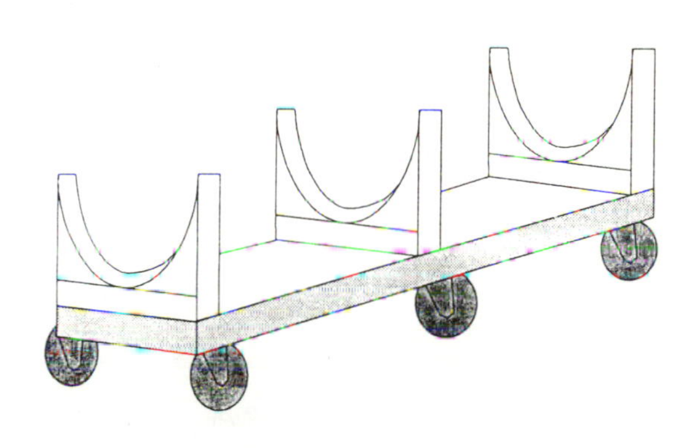

Figure 13.19 Bar cradle truck.

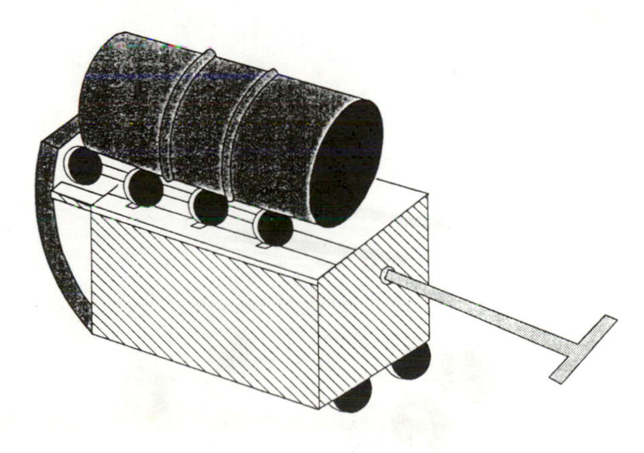

Figure 13.20 Portable drum rotator.

Figure 13.21 Pallet jack.

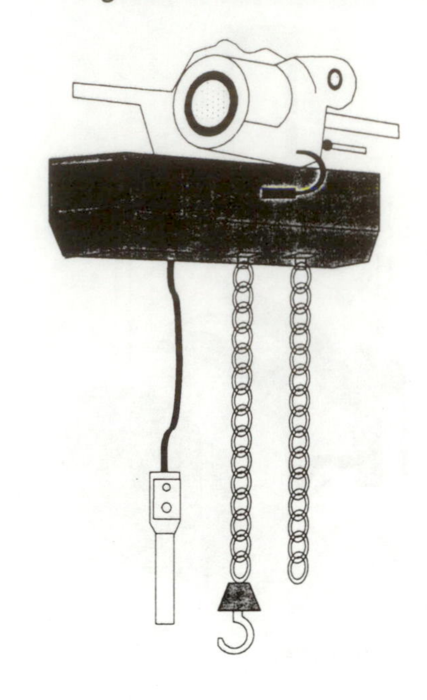

Figure 13.23 Chain hoist.

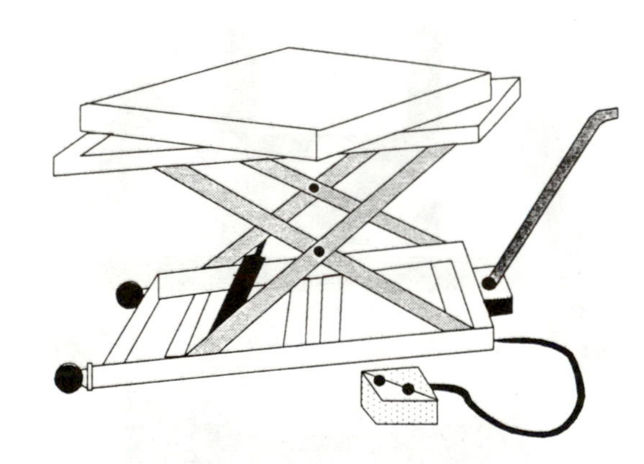

Figure 13.24 Rotating top lift table.

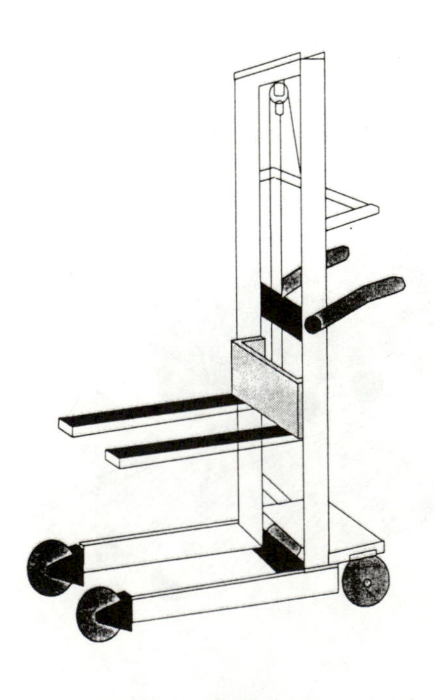

Figure 13.22 Mobile elevator.

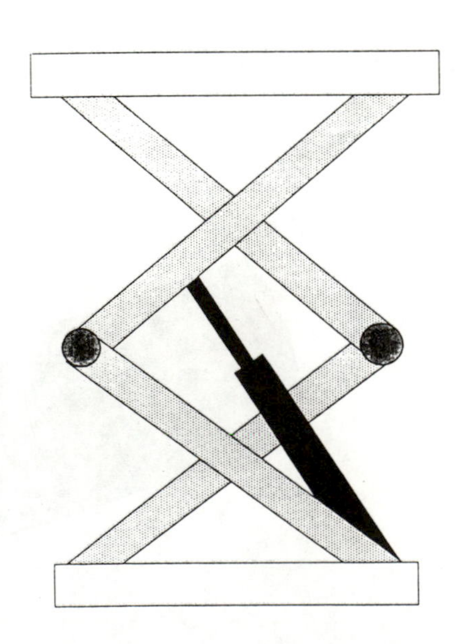

Figure 13.25 Double scissors lift table.

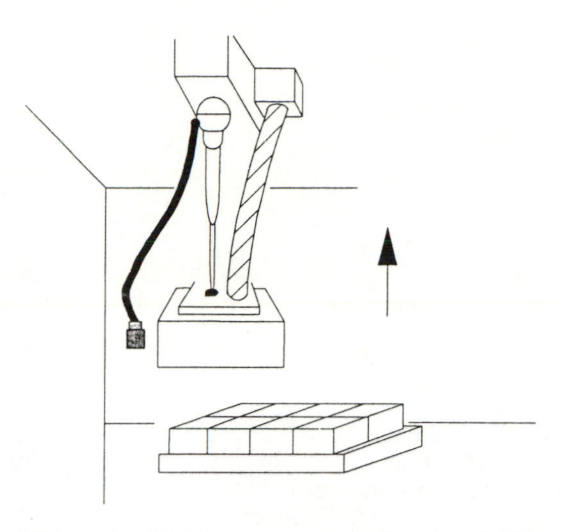

Figure 13.26 Vacuum type below-hook lifter.

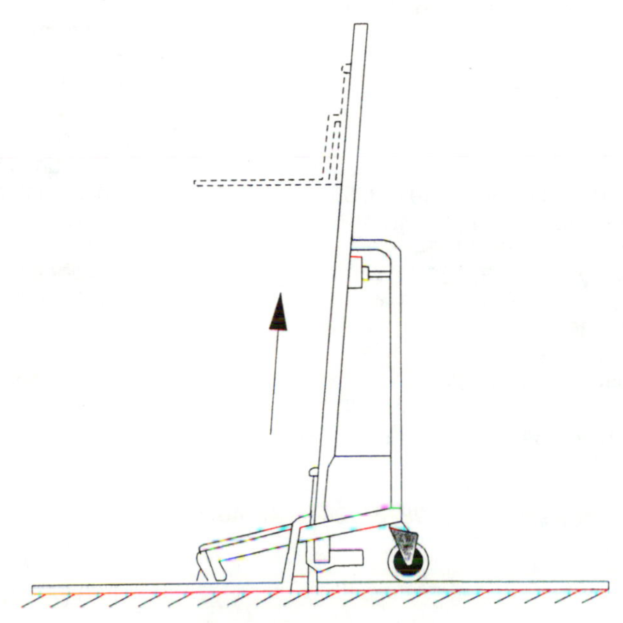

Figure 13.27 Walkie stacker.

Figure 13.28 Furniture dolly.

Index

acceleration effect 53
acclimatization (hot environments) 24
action limit 5, 47, 49
aerobic capacity 16, 39–41, 43, 121
  rest allowances 123–7 *passim*
age (risk factor) 15–16
American Heart Association 124
anthropometry 14–15, 16, 21, 85
Association d'Assurance (Luxembourg) 4
asymmetrical lifting 22–3, 29, 36–7, 67, 68, 86, 113
asymmetrical loads 22–3, 67–8, 85, 86
Australia (advisory standards) 5–6, 7

back
  margin of safety for 48–55, 61
  spinal compression 30–1, 33–8, 47–55
  spinal mobility 15, 16
back injury 4–5, 19
  biomechanical studies 34–8, 49–53, 55
  epidemiological studies 28–34, 48–9, 55
  physiological studies 38–41, 54–5
  previous history of 15, 18, 33
  psychophysical studies 41–3, 53–4
bar cradle truck 135
barrel carrier, tilting 134
basal metabolism 122, 123, 124
below-hook lifter 132
  lifting-beam type 134
  vacuum type 137
bicycle aerobic capacity 40–1
biomechanical models 5, 16, 25, 28, 31, 34–8, 42, 46–55
body size, strength and 14–15
box hauling device 133

carrying
  load transfer 116, 118, 119–20
  one-handed 86
  two-handed 84–6
case studies (job design/evaluation) 108–15
centre of gravity movement 22–3, 68
chain hoist 136
clothing, protective 22, 24
coefficient of friction 22
compressive force (spinal) 30–1, 33–8, 47–55, 61
conveyor 133
cooling jackets 24
core temperature 24
cost of injuries 4, 5
counts (in epidemiology) 29
couplings 5, 11, 21–2, 68, 85, 86
crane, portable 132
cumulative injury rate 31–2
cut-out handles 21–2
cylindrical handles 22

delay allowances (rest) 122
design/evaluation procedure 107–15
design approaches
  biomechanical 28, 34–8, 49–53, 55
  comparisons 42–55
  epidemiological 28–34, 48–9, 55
  physiological 28, 38–41, 54–5
  psychophysical 28, 41–3, 53–4
design database (for industrial workers) 62–9
dollies 134, 137
double scissors lift table 136
drum
  lifter 135
  portable rotator 135
  truck and drainer 134
  upender 135
  vertical lifter 135
dynamic strength testing 18

electromyographic activity 22, 53, 55
elevator, mobile 136
elevator, portable 133
employers, checklist for (HSC) 11
End-of-Train (EOT) device 113–15
endurance time 22, 120, 121
epidemiological approach 28–34, 48–9, 55
equipment, protective 24
European Coal and Steel Community guidelines 7, 12–13
European Community directives 10–11, 12
evaluation procedure (multiple-task MMH) 107–15

fatigue allowances (rest) 122
feet (positioning) 22
flooring materials 22
force limit 7, 12–13, 20
four-wheel hand truck 133
free-style posture 19
frequency of handling 5, 10, 22
frequency of lifting 30, 33, 38, 41–2, 48, 61, 63–6, 70, 116–21
furniture dolly 137

gender differences (risk factor) 16
goggles 24

hand truck, four-wheel 133
handles 21–2
handling, repetitive 22, 37, 38
handling techniques 19–20
hazards (safety aspects) 23
Health and Safety Commission 3–4, 6, 11
Health and Safety Executive 7
heart rate 117, 118, 119
heat stress 24, 69, 86
high frequency manual handling tasks 116–21
history-taking (importance of) 33
hoist, chain 136
holding 19, 87–90
hot environments 24
hydraulic dumping table 133

incidence rate 29
inertia effect 53
injuries
  back *see* back injury
  musculoskeletal 3, 4, 16, 107
  overexertion 3, 5, 15, 28–9, 116, 122
  statistics 3–5
International Labour Office 6, 10

intervertebral disc L5/S1 30–1, 34–6
intra-abdominal pressure 16, 19, 22, 25, 28, 34–5, 38, 47–8, 53, 55, 70, 74
isointertial strength 18
isokinetic strength 18, 49, 50, 51
isometric pulling force 46, 82
isometric push force 46, 74, 102, 104–6
isometric strength 18, 22, 47, 49, 50, 54, 91

jack, pallet 136
Japan (legislation) 7, 8, 9, 11
job design/evaluation (multiple task MMH) 107–15
job rotation 25
Job Stress/Severity Index (JSI) 31–2, 33–4, 36, 49, 61

kneeling hold 87, 90
kneeling lift 92, 94, 95, 97, 98

L5/S1 disc 30–1, 34, 35–6
legislation 5–13
leisure metabolism 122, 124
Liberty Mutual Insurance Company 48–9
lift table 133, 136
lifting
- asymmetrical 22–3, 29, 36–7, 67, 68, 86, 113
- frequency 30, 33, 38, 41, 42, 48, 61, 63–6, 70, 116–19
- in limited headroom 23, 67, 69
- one-handed 42, 69–71, 97–101
- symmetrical 37–8, 62, 63–6
- two-handed 42, 61–9, 91–6, 99–101
- two-person 71
- unusual postures 91–101
- work duration and 62, 67

lifting-beam type below-hook lifter 134
lifting strength ratio 30–1
limited headroom 23, 67, 86
load
- asymmetry 22–3, 67–8, 85, 86
- characteristics 20–1
- placement clearance 68
- shape 20
- size 20–1, 85
- transfer (weight limits) 116–20

lowering (weight limits) 116–19
lumbar spine
- compressive strength 33, 35–8, 48, 50, 51, 53, 54, 61
- mobility 15, 16

Manual Materials Handling (Ayoub and Mital) 14, 39
masks 24
maximum acceptable frequency of lift 42
maximum acceptable weight of carry 119
maximum acceptable weight of lift 42, 47, 116–19
maximum permissible limit 5, 47, 49–50
maximum voluntary contraction 49–52, 54
mechanical aids 132–7
metabolic energy expenditure 15, 19, 22, 24, 38–43, 47, 54, 62, 122–3
mobile elevator 136
motivation 17
multiple-task jobs (design/evaluation procedure) 107–15
muscular contraction 16, 122
muscular strength 16–18, 21
musculoskeletal injuries 3, 4, 16, 107
musculoskeletal system 25, 28, 29

National Board of Occupational Safety and Health (Sweden) 3
National Institute of Occupational Safety and Health (USA) 5–6, 7, 47, 49, 50, 107, 127
National Safety Council (USA) 4

Netherlands (legislation) 7, 11–12

one-handed carrying 86
one-handed lifting 69–71
one-handed pulling 78, 83
one-handed pushing 74, 77
osteoligamentous preparation 52
overexertion 3, 5, 15, 28–9, 116, 122
oxygen consumption 39–40, 117, 118, 119
oxygen transport system 38
oxygen uptake 43, 44

pallet dolly 134
pallet jack 136
personal allowances (rest) 122
personal risk factors 29, 33
physical fitness 15
physical training 17–18
physiological approach 5, 28, 29, 38–43, 45–7, 48, 54–5
physique 14–15
platform truck 133
portable crane 132
portable drum rotator 135
portable elevator 133
posture
- fixed 20, 22, 25
- handling techniques and 19–20
- unusual 91–106

prevalence rate 29
program output (rest allowance) 125–7
program source listing (rest allowance determination) 128–31
protective equipment/clothing 22, 24
psychophysical approach 5, 17, 28, 41–3, 45–8, 53–4
pulling 20
- one-handed 78, 83
- two-handed 78–82
- unusual postures 103, 106

pushing 20
- one-handed 74, 77
- two-handed 72–3, 74, 75–6
- unusual postures 102–6

ratings of perceived exertion 116–19
Recommended Dietary Allowances 124
Recommended Weight Limit 5, 42–8, 62–6
repetitive handling 22, 37, 38
rest allowances 24, 25
- determination procedure 122–6
- determination program (source listing) 128–31
- examples (program output) 125–7
- types of 122

risk factors 14–25, 29–30, 33, 39
risk potential 108–15 *passim*
rotating top lift table 136

safe limits 5
safety aspects (hazard list) 23
safety margin (for the back) 48–55, 61
sagittal plane 21, 22–3, 70, 85, 86
screening 18, 107
seated posture 19–20
sedentary work 19
selection 18
semi-squat posture 19
shear force 22, 34, 35, 37–8
shelf opening clearance 23
shoe soles 22, 24
side lift 96, 98, 99
simulated job dynamic strengths 18
sitting hold 87, 90

sitting lift 69–70, 71, 92–3, 95, 97
skid truck 134
software (rest allowance calculation) 128–31
spatial restraints 23, 25, 67
spinal compression 30–1, 33–8, 47–55, 61
spinal mobility 15, 16
squat posture 19, 20
squatting hold 87, 90
squatting lift 92–3, 95, 97
squatting push 102, 103
stacker, walkie 137
standing hold 87, 90
standing lift 69–70, 91, 93, 94, 97
standing push 102, 103
standing vertical reach 87, 90
State Electricity Commission of Victoria (Australia) 4
static work (effects) 19
statistics/data 3–5
stoop posture 19, 20
strength 14–15, 22, 47, 49–51, 54, 91
  testing 18
symmetrical lifting 37–8, 62, 63–6
symmetrical loads 20, 22, 84–5

task duration 24–5, 62, 67
tilting barrel carrier 134
training 17–18
treadmill aerobic capacity 40–1
tripod dolly 134
truck, bar cradle 135
truck, drum (and drainer) 134
truck, four-wheel hand 133
truck, platform 133
truck, skid 134
truck, utility 135
turning 19, 33, 118, 119–20
two-handed carrying 84–6
two-handed lifting 42, 61–9, 91–6, 99–101
two-handed pulling 78–82
two-handed pushing 72–3, 74, 75–6
two-person lifting 71

UK (legislation) 6, 7, 10, 11
USA (guidelines) 5, 7
utility truck 135

vacuum type below-hook lifter 137
vertical drum lifter 135
very high frequency MMH tasks 116–21

walkie stacker 137
weight/force (acceptable) 41–2, 108
weight limits 5, 7–10, 24–5
  high/very high frequency tasks 116–20
  recommended 5, 42–8, 62–6
  unusual postures 99–101
wet bulb globe temperature (WBGT) 24, 69
work organization 25
Work Practices Guide to Manual Lifting 5, 6, 7, 49
work rate 108–12 *passim*
working duration 62, 67, 86, 122
workplace geometry 25
workplace layouts 107, 108–9, 111
workplace risk factors 29, 33, 39
workstation organization 10–11, 12